Bill Manto

Electronics

In easy steps is an imprint of In Easy Steps Limited
16 Hamilton Terrace · Holly Walk · Leamington Spa
Warwickshire · United Kingdom · CV32 4LY
www.ineasysteps.com

Acknowledgements
We thank www.upsbatterycenter.com and chem.libretexts.org for allowing use of their dry cell and lead-acid cell images, respectively.

In Easy Steps Limited supports The Forest Stewardship Council (FSC), the leading international forest certification organization. All our titles that are printed on Greenpeace approved FSC certified paper carry the FSC logo.

Printed and bound in the United Kingdom

ISBN 978-1-84078-759-7

Contents

1 Basic Principles

This chapter is a simple introduction to electricity, what it is and how it is produced. You will learn the standard units and symbols used in electronics and be shown the basic components.

Introduction

Developments in electronics over the years have had a profound effect on the way we lead our lives. We are so surrounded by electronic gadgets that if electricity had never been discovered it's almost impossible to imagine what our lives would now be like.

The result of numerous electrical discoveries and developments over the centuries means that we have today become more technologically aware than ever. Consequently, you will be reading this book because you want to learn more about electronics, components, and circuits.

To understand all of this, first you have to learn, in simple terms, a little bit about electricity and how it works because that is the power behind electronics – this is what this chapter shows you.

Don't forget

The world has become totally reliant on electricity as a prime energy source. Just think where we would be today if it had never been discovered.

The invisible force

Electricity was not invented but discovered. It is a form of energy that occurs in nature and so was always there, but we couldn't easily see it other than, for example, during a lightning storm. It was realized quite quickly that there existed some sort of strong, invisible energy or force that deserved investigating further.

Centuries of experimenting and development has given us the basic electrical laws and principles that now allow us to not only understand electricity, but to use it to create electronic circuits and make electricity work for us.

Here are three of those credited with some of the early work:

- **Benjamin Franklin** – established the connection between lightning and electricity.
- **Alessandro Volta** – discovered that certain chemical reactions could produce electricity, and created a crude electric battery.
- **Michael Faraday** – discovered a mechanical method of generating electricity when he created the electric dynamo.

So, electricity wasn't discovered by just one person. The concept of an invisible energy had been known about for thousands of years, but it was a combination of the individual efforts of many great minds that has given us the understanding we have today.

Although electricity can be fun to experiment with, it can also be dangerous when working with high voltages. A mistake could be fatal if the correct precautions are ignored, so always take note of any safety advice or warnings.

Electrical Terms

Energy can take many forms; like heat energy, for example. Electricity is simply the name given to electrical energy and, as with heat, there needs to be a difference in potential for it to travel from one point to another. But how does electricity travel?

All matter is made up of *atoms* bonded together in some way. At the center of the atom is the *nucleus*, made up of positively charged *protons* and *neutrons* (which have no charge). Negatively charged *electrons* orbit around the nucleus.

Basically, if the number of electrons and protons is equal then the atom is said to be stable and has no charge. Atoms can be made unstable by rubbing two materials together so that electrons transfer from one material to the other, leaving the atoms effectively with a positive or negative charge. You can see this *electrostatic* effect when pulling an item of clothing on or off over your head and it crackles or causes your hair to stand on end!

This movement of electrons from one point to another is seen as the flow of electricity. Below is some electrical terminology:

Charge (C)

All protons and electrons have a tiny amount of electrostatic charge. This charge is measured in *coulombs* (Q).

Current (I)

This is the movement of electrons around an electrical circuit and is defined as the rate of flow of charge. Its unit is the *ampere* (A).

Voltage (V)

Electromotive force (emf) is what creates the flow of current in a circuit and is measured in *volts*. The potential difference (pd) is the voltage difference or voltage drop between any two points.

Power (P)

This is a measure of the rate at which energy is transferred. Power is measured in *watts* (W).

Conductor

A material with lots of charge-carrying free electrons, such as metal.

Insulator

A material where the electrons are firmly bound to the nucleus of its atoms so that they cannot move and hence conduct charge.

Where an atom has more electrons than protons it is negatively charged, and positively charged if it has fewer electrons than protons.

One ampere of current is calculated by:

$I = Q/t$

(**t** is the time in seconds and **Q** is the charge.)

The effects of the flow of an electric current can be detected in many different ways – for example, as heat, light, magnetism, etc.

Primary Cells

Electrical energy can be produced by a number of means, including mechanical and chemical. A device that generates a charge when a chemical reaction takes place is called a *cell*. There are two main types: primary and secondary. First, we look at primary cells.

Hot tip

In 1799, a year after discovering methane, the noted pioneer of electricity and power, Alessandro Volta, proved that electricity can be produced by chemical means when he invented the voltaic pile: a crude form of battery.

The simplest primary or *voltaic* cell consists of the following:

1. A positive electrode (anode) consisting of a copper plate
2. A negative electrode (cathode) consisting of a zinc plate
3. An electrolyte of dilute sulfuric acid

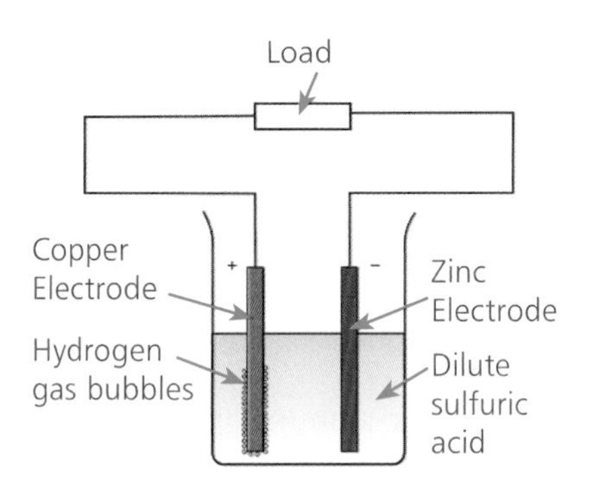

The sulfuric acid is poured into a container, and the electrodes are placed into the electrolyte. If the two electrodes are then connected together outside of the cell, a current will flow from the copper electrode to the zinc electrode, and through the electrolyte back to the copper electrode.

Connecting several cells together forms what is called a *battery*.

In this simple form, the voltaic cell only works properly for a short time. As it generates current, a layer of hydrogen bubbles starts to build up on the copper electrode, causing its output to become less and less. Also, the zinc electrode has to be totally pure. If not, any impurities will react with the zinc and the sulfuric acid, again reducing the cell's output.

Dry cell

The energy produced is called *electromotive force* (*emf*). A typical simple primary cell described above has an emf of about 1 volt. More common is the *dry cell*, as used in torches, producing about 1.5 V. This is an electric cell in which the electrolyte is in the form of a paste to prevent any spillage.

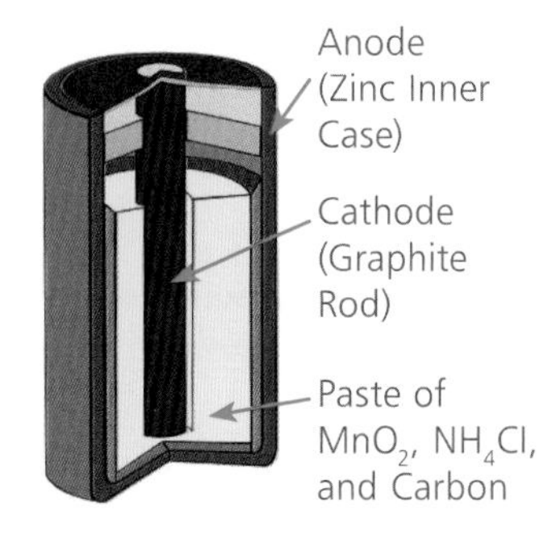

Primary cells use up the chemicals they contain. They cannot be recharged, as the action is non-reversible.

Secondary Cells

The lead-acid cell

This is one of the most common secondary cells and does not, on its own, generate electricity. The cell has to be initially charged with electrical energy from an external source; this energy then being stored in the cell as chemical energy.

Secondary cells are reusable in that they can be charged from an external source and then discharged during use many times over.

1. The construction of the lead-acid cell is quite complex

2. Cells contain special interlaced positive and negative plates

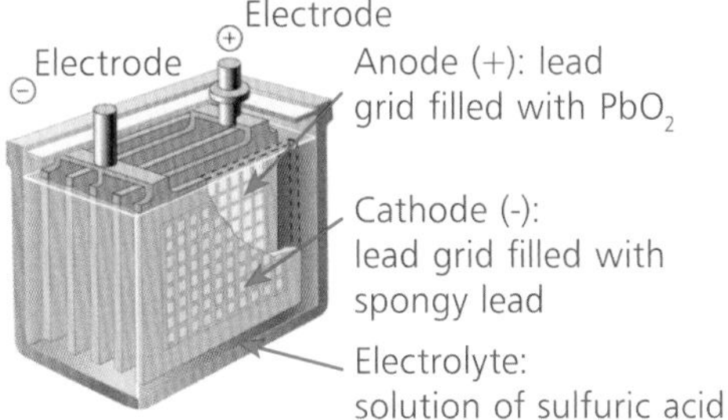
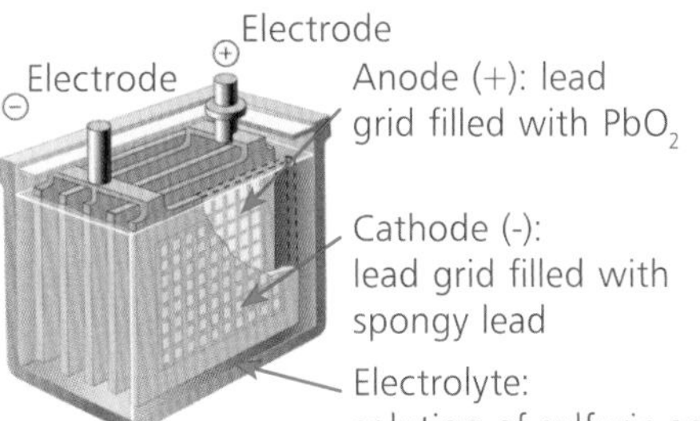

3. The electrolyte is a mixture of sulfuric acid and water

4. A charged lead-acid cell has an emf of approximately 2.0 V when in use

Small rechargeable batteries in everyday use contain a chemical paste or *solid* electrolyte instead of sulfuric acid. Typical types include:

- Ni-Cd (Nickel Cadmium)
- Ni-MH (Nickel Metal Hydride)
- Li-ion (Lithium Ion)

The advantage of the lead-acid cell is that the charging process is reversible so that once charged, the chemical energy can be released in the form of an electric current as required.

When the cell has released all of the stored energy and has become discharged, then it can simply be recharged again from the external source and the process repeated.

You will actually be more familiar with this type of cell than you think. A car battery is, in fact, a number of lead-acid cells joined together. It stores and provides the electrical energy required by the motor vehicle, and is recharged when the engine is running.

The voltage rating of a Ni-Cd or Ni-MH rechargeable battery is usually a little lower than that of its dry cell equivalent. They should not be mixed in use!

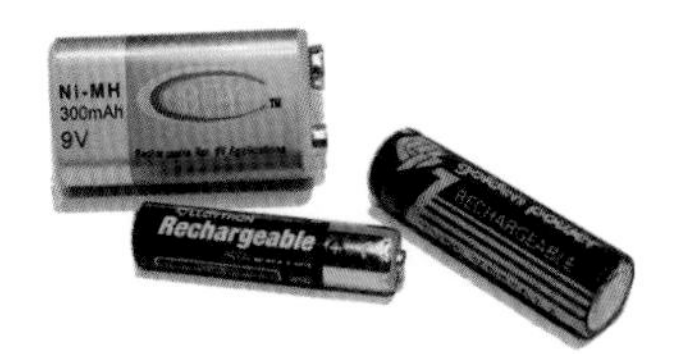

Although not lead-acid, here are some popular everyday types of small rechargeable secondary cells you will recognize.

Electrical Units

Before going any further, now is the time for an introduction to the units associated with electrical quantities and their symbols, where they come from, and where they are used. Knowing this important information will make it easier for you to understand electronic principles and circuit diagrams as we work through them. Some you will already know, but others will be new to you.

Basic SI units can be combined to form derived SI units – for example, meters per second (m/s).

SI units

Engineers use the International System of Units (Système Internationale d'Unités), which is metric based and usually simply called *SI units*. This system covers not only electrical units but other familiar units, as you can see from these few examples:

Quantity	Unit	Meaning
Electric Current	A	ampere
Time	s	second
Mass	kg	kilogram
Length	m	meter

The SI units can be made to represent smaller or larger quantities by using a prefix. This signifies how much to multiply or divide the value by:

Prefix	Name	Meaning
M	mega	multiply by 1,000,000 (i.e. value x 10^6)
k	kilo	multiply by 1,000 (i.e. value x 10^3)
m	milli	divide by 1,000 (i.e. value x 10^{-3})
μ	micro	divide by 1,000,000 (i.e. value x 10^{-6})
n	nano	divide by 1,000,000,000 (i.e. value x 10^{-9})
p	pico	divide by 1,000,000,000,000 (i.e. value x 10^{-12})

The prefix is case-sensitive so it is important to avoid mistakes by always using the correct upper or lower case letter.

For example, MV would mean megavolt; mA would mean milliamp; and μA would mean microamp.

Work (or energy)

Because electricity is a form of energy, this energy is measured using the standard unit of work or energy, the *joule* (J).

You may remember Ohm's law from school. Although you probably thought you would never use it, it will now be very handy for simple calculations you may have to make – for example, establishing the current drawn by a circuit you have built so you can protect it with the correct value fuse.

Power

The unit of power is the *watt* (W), and the symbol for power is P. One watt is equal to one joule per second and is calculated by:

power $(P) = W/t$ therefore, energy $(W) = Pt$

(Where W is energy or the work done in joules, P is the power in watts and t is the time in seconds.)

Charge

The unit of charge is the *coulomb* (C), and the symbol for charge is Q. One coulomb is equal to one ampere second:

charge $(Q) = It$

(Where Q is the charge in coulombs, I is the current in amperes, and t is the time in seconds.)

Electrical potential (and emf)

The difference in potential between two points in a conductor or electric circuit is called *electrical potential.* A change in that electrical potential is called a *potential difference*. The unit of electrical potential is the *volt* (V), and the symbol is V. One volt is equal to one joule per coulomb. Voltage is calculated as follows:

volts $(V) = P/I$ therefore, power $(P) = IV$

(Where V is the voltage in volts, P is the power in watts, and I is the current in amperes.)

Electromotive force (emf), symbol E, is also measured in volts.

Try not to become confused by all the symbols and abbreviations used. Take some time to study these few pages and all will become clearer later.

Resistance

Opposition to the flow of electrical current is called *resistance*. Its unit is the *ohm* (Ω), and the symbol for electrical resistance is R. As one ohm equals one volt per ampere, resistance is calculated by:

resistance $(R) = V/I$ therefore, $V = IR$ and $I = V/R$

(Where R is the resistance in ohms, V is the potential difference in volts across the resistance and I is the current in amperes flowing through the resistance. The above is called *Ohm's law.*)

...cont'd

Capacitance

When a voltage is applied across two parallel conducting plates separated from each other by air, for example, an electric field and hence an electric charge builds up in the area between the plates.

Capacitance is the term used to indicate how much charge can be stored between the plates for a given voltage. The unit of capacitance is the *farad* (F), and its symbol is C. It is calculated by:

$$\text{capacitance } (C) = Q/V \qquad \text{therefore, } Q = CV$$

(Where C is the capacitance in farads, Q is the charge in coulombs and V is the potential difference in volts between the plates. Note that typical capacitance values are in the order of μF, nF or pF.)

Capacitors are covered in more detail in Chapters 2 and 3.

Don't forget

A capacitor is very much like a secondary cell in that it will hold a charge when a voltage is applied across it, but as this charge is small, it will release it very quickly compared with a cell.

Summary of SI units and symbols

It is important to remember that there is a difference between a unit and a symbol. The unit expresses a value, whilst the symbol is just that: a symbol in a formula. Do not get the two mixed up.

Also, do not confuse the symbols, as they may not always be obvious. For example, it's easy to remember C is for capacitance but then it can't be used again for current, so current is allocated the symbol I, which appears unrelated! Use the table below to help you remember:

Quantity	Symbol	Unit Name	Unit
Work (or Energy)	W (or E)	joule	J
Power	P	watt	W
Charge	Q	coulomb	C
Potential Difference	V	volt	V
Electromotive Force	E	volt	V
Resistance	R	ohm	Ω
Capacitance	C	farad	F
Time	t	second	s
Mass	m	kilogram	kg
Length	l	meter	m

Do not confuse units with symbols and always remember that both upper case and lower case characters have specific meanings.

Insulators and Conductors

Another important element of understanding electricity is knowing what will conduct it and what won't, because all electronic components are based on these basics. To do this we take a quick look at what is meant by the *flow of electric current*.

Copper and gold are excellent conductors, and so are used extensively in electronic circuits. You will often come across gold-plated pins in high-quality connectors.

Atoms and electrons

In simple terms, everything is made up of *atoms* that consist of *protons*, *neutrons,* and *electrons*. The protons have a positive electrical charge and together with the neutrons, which have no charge, make up the nucleus of the atom. Outside of the nucleus are the tiny negatively charged particles we call electrons.

The atoms of different materials all have different numbers of these protons, neutrons, and electrons. A powerful force keeps everything bonded together and balanced, but it is also possible for an atom to "lose" an electron. The atom is now electrically unbalanced and has a positive charge, making it able to attract an electron from another atom.

This movement of electrons is normally random, but if a voltage is applied across the material then the electrons all move in the same direction. This movement is the flow of electric current. Some materials conduct electricity very well, whilst others conduct hardly any or none at all.

Insulators

Materials where the electrons are held tightly to their nucleus so that they exhibit hardly any current flow are called *insulators*. They have a high resistance to the flow of current.

Conductors

Materials that have loosely connected electrons to their nucleus are called *conductors* because these loose electrons are able to easily move from one atom to another through the material. They have a low resistance to the flow of current. Metals are in this category.

Wherever there is electricity it is always advisable to work with insulated tools to prevent shorting out connectors or components. It will also minimize the chance of receiving an electric shock if high voltages are present.

The table opposite lists some common insulators and conductors that you will be familiar with.

Insulators	Conductors
Rubber	Copper
PVC	Aluminum
Glass	Mild Steel
Wood	Gold

Circuit Components

With some of the electrical theory out of the way, you are now ready to take a look at the basic electronic components that are used in circuits. There are quite a few different ones, but the following are the most common ones that you will come across.

Resistors come in all shapes and sizes depending on their application. In general, small resistors are for very low wattage applications, whilst larger resistors are often used in high-current applications where they may get hot and so have to dissipate the heat.

Resistors

An electronic circuit will contain lots of resistors. They are used to limit the flow of current and are available in all sorts of shapes and sizes and a wide range of values and tolerances. There are also different types such as carbon, metal film, and wirewound. Which type is used depends on its specific purpose in the circuit.

The resistor color code

Because resistors are often very small it would be difficult to mark their value and tolerance on the casing. Such markings might also become burned off after prolonged use, as circuits generate heat. Instead, small resistors are commonly identified by colored bands. The four-band and the five-band resistor color codes are shown below, together with an example of each.

4 Band Code — for 2%, 5%, 10% — = 260kΩ ± 5%

COLOR	1st BAND	2nd BAND	3rd BAND	MULTIPLIER	TOLERANCE
Black	0	0	0	1Ω	
Brown	1	1	1	10Ω	± 1%
Red	2	2	2	100Ω	± 2%
Orange	3	3	3	1kΩ	
Yellow	4	4	4	10kΩ	
Green	5	5	5	100kΩ	± 0.5%
Blue	6	6	6	1MΩ	± 0.25%
Violet	7	7	7	10MΩ	± 0.10%
Grey	8	8	8		± 0.05%
White	9	9	9		
Gold				0.1	± 5%
Silver				0.01	± 10%

5 Band Code — = 537Ω ± 1% — for 0.1%, 0.25%, 0.5%, 1%

The more you use the resistor color code table, the easier it becomes to remember the bands and their values.

The four-band is the most common as it is used for wide tolerance (and hence cheaper) resistors found in many circuits. The five-band code is used where very accurate resistance values are needed.

Note that the bands for the value are grouped together on the left, whilst the tolerance color is spaced apart at the other end. To read the value of a resistor with four colored bands just do this:

1. With the resistor the right way round, note the first color and look up the value in the 1st band column (e.g. Red = 2)
2. Note the second color and look up its value in the 2nd band column (e.g. Blue = 6, so the significant value is 26)
3. Note the third color, look up its value in the Multiplier band column and multiply the significant value with this figure (e.g. Yellow = 10 kΩ, so 26 x 10 kΩ = 260 kΩ)
4. Note the fourth band color and look its value up in the Tolerance band column to get the full resistor value (e.g. Gold = ±5%, giving the resistor's value of 260 kΩ ±5%)

Always check the value of a resistor with a multimeter set to the Ohm's range before using it in a circuit. Humans can make mistakes when reading a color, as can a manufacturer when labeling a resistor!

Capacitors

A capacitor is a device for storing energy in the form of an electrical charge, like a battery or cell, but stores a much smaller charge and can be charged or discharged almost instantly. Capacitors have many uses, such as in power supplies or tuned circuits.

Take care when working with large-value capacitors in case they are still charged. Always ensure they are fully discharged before attempting any work in that part of a circuit. The last thing you want is for the stored energy to discharge through another component with disastrous results, or worse, through YOU!

Diodes

A diode is an electronic device that allows current to flow in one direction only, like a one-way valve but with no moving parts. There are various types of diodes with specific functions such as controlling a voltage level or emitting a colored light (LED).

Power sources

Electronic circuits require a power source to provide the electrical energy for them to work. Batteries are often used to provide a simple power source, but when carrying out repairs or practical experiments it is more common to use a stabilized power supply.

This type of mains-driven power supply provides an accurate voltage that can be set by the user. There is often a means of limiting the current that the unit will supply, and protection circuitry to effectively turn off its output if the circuit it is powering malfunctions.

Standard Symbols

Beware

Old habits die hard with engineers. Many still hand draw circuits using the old symbol for a resistor – a sawtooth line like this:

Circuit diagrams are the drawings used to show how components are connected together in an electrical or electronic circuit. For standardization, symbols are used to represent the components. This means that it should be possible to read and understand a circuit diagram created anywhere in the world.

Common standard component symbols

The following are some of the more common standard symbols used in electrical and electronic circuit diagrams that you need to become familiar with. Many of these will be covered in more detail as you work through this book.

Symbol	Meaning	Symbol	Meaning
	Cell		Wiring Connections
	Battery		Earth/Ground
	Battery (alternative)		Fuse
	Resistor	V	Voltmeter
	Variable Resistor	A	Ammeter
	Capacitor		Diode
+	Capacitor (polarized)		Zener Diode
	Variable Capacitor		Light Emitting Diode
	Switch	B C E	NPN Transistor
	Lamp	B C E	PNP Transistor
	Inductor		Transformer

Although we have standard symbols, you will often see "deviations" from these standards. For example, you may come across a diode drawn with a white body, not black, and the center wire going right through it. Don't be confused – the symbol will be close enough to the standard symbol for you to recognize which component it represents.

The above table only shows a selection of basic symbols; they may vary slightly depending on the actual component or by country. However, it is usually possible to easily interpret these variations.

2 DC Circuits

Learn about the types of DC circuits and how current flows in them. You are then introduced to Ohm's law and shown how to use it to perform simple calculations for voltage, current, and resistance, followed by circuits involving resistors and capacitors in series and parallel configuration.

Characteristics

A circuit is the term used to identify how various components such as resistors, capacitors, diodes, and transistors are connected together to perform a particular function.

When the circuit is powered by a battery or stabilized power supply it is called a *DC circuit* because the supply voltage is at a fixed constant level. This voltage is always at a positive potential relative to zero volts (normally referred to as *ground* or *earth*). If the voltage changed from positive to negative values relative to zero volts then it would be called an *AC circuit* because of the voltage source's alternating nature.

DC stands for Direct Current; the type of voltage source provided by a battery.

AC stands for Alternating Current. This is where the source is changing from a positive to a negative value at a fixed rate. An example of an AC source is the domestic mains supply.

It is easier to start with DC circuits first because they are simpler to understand. You will learn more about AC circuits in Chapter 5.

The circuit diagram

A very simple circuit can be constructed using just three things – a lamp, a switch, and a battery or cell to provide power. To show how they are wired together, a diagram is used. However, to draw the actual components and the connecting wires would be very time consuming, especially if it was a large circuit, so instead the circuit is drawn using standard component symbols, such as those on page 18 in Chapter 1. Drawing a circuit diagram using standard symbols also means it can be understood around the world.

A simple DC circuit

The diagram opposite represents the basic DC circuit mentioned above. It shows how the component symbols for a switch, a lamp, and a battery are connected. At the moment no current flows because the switch is open, but when closed to complete the circuit, current can flow and the lamp will light. This is how you "read" a circuit diagram.

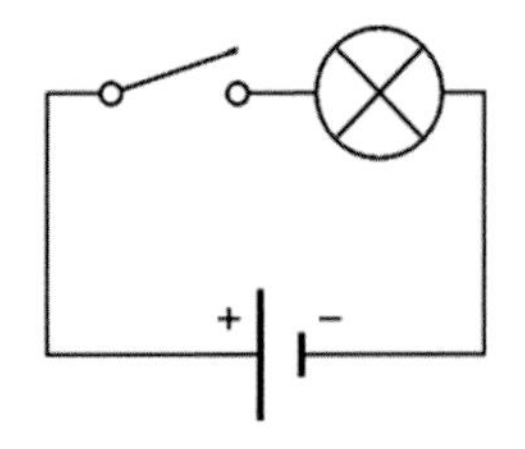

Conventional current flow originates from when electricity was first discovered and electrons were unknown. It was assumed that current flowed from positive to negative, but as an electron is negatively charged, and unlike charges attract, we now know that the true flow of electrons is from negative to positive!

Current flow

Current is defined as the flow of electrons, and by convention is said to flow from positive to negative. The opposition to the flow of current is called *resistance*, and in a typical circuit with many components there will be many paths for the current to flow along, and various resistances to this flow, as we shall see later. In our simple circuit, power is dissipated as light and heat by the lamp, plus a magnetic field is present in the wires carrying the current.

Current

From Chapter 1 you will remember that current is measured in amperes, normally shortened to just *amps*, and the symbol is A. To measure current, a meter called an *ammeter* is used. As this is placed in series in the circuit so that the current flows through the meter, it is important that it has as low a resistance as possible so as not to oppose the flow of current in any way and hence give a wrong reading. The multimeter is a popular and handy tool for doing this because (as the name suggests) you can select different functions on it, including ammeter. This setting will have a range of current levels that it can measure – from very small to large.

The current setting on a multimeter is often protected by an internal fuse, just in case a high current is being measured without first selecting the appropriate range. If the ammeter setting doesn't appear to be functioning correctly, check the fuse.

Although it can sometimes be high, the current in most electronic circuits will be small. When lower than 1 A it is normal to express it in sub-units using the milliamp or microamp symbol.

A milliamp is equal to one-thousandth of an amp:

$1 \text{ mA} = 0.001 \text{ A} \quad (\text{or } 1 \times 10^{-3} \text{ A})$

A microamp is equal to one-millionth of an amp:

$1 \mu\text{A} = 0.000\,001 \text{ A} \quad (\text{or } 1 \times 10^{-6} \text{ A})$

Current in a series circuit

The circuit opposite now shows two lamps connected in series. If the current I were measured as it leaves the positive battery terminal, in the lead between the two lamps and in the negative lead back to the battery, it would always be the same reading. Thus, the current at any point in a series circuit is always the same.

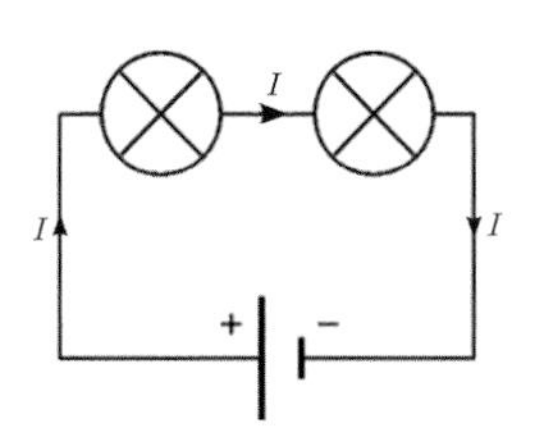

The current flowing back into the power source will always be equal to the current flowing out of the power source; no current is lost as it makes its way through a circuit, irrespective of the type of circuit.

Current in a parallel circuit

If the lamps are instead connected across each other as per the second circuit, they are said to be *in parallel*. The measured current (I) leaving and returning to the battery would be the same, but the current flowing into each lamp (I_1 and I_2) would be a different and smaller amount; the actual value depending on the resistance of each lamp. These two readings added together will equal the total current returning to the battery ($I = I_1 + I_2$).

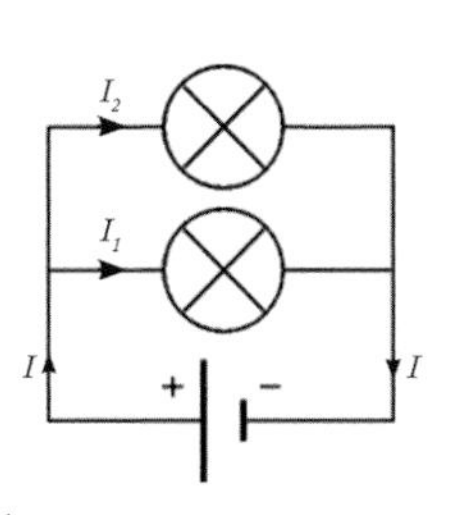

Voltage

The force that drives a current around a circuit is called *voltage* and is provided by the circuit power source, such as a power supply unit or a battery. This is why voltage is also known as the *electromotive force* or *emf*. The unit of voltage is the *volt* and the symbol is V.

You can connect power sources such as batteries or cells in series to increase the voltage supply. The total voltage supplied will be the sum of the voltage of each individual battery or cell.

When referring to voltages developed in DC circuits, the term *potential difference* is often used, meaning the difference in electric potential between two points. This potential difference is measured by placing a voltmeter across the two points in question. For example, placing the two leads of the voltmeter across a resistor in a circuit will give the potential difference, or voltage, across that resistor (provided the power is switched on, of course).

As with measuring current, a multimeter can be used for this once set to the appropriate voltage range. However, to measure voltage the meter is placed in parallel across the component and not in series with it as for current.

To increase the current available from a power source, simply connect batteries or cells in parallel. The total current available will be the sum of the current of each individual battery or cell.

Voltage in a series circuit

In a series circuit the sum of the voltage measured across each component will always be equal to the value of the supply voltage. Using the circuit opposite, if the voltage is measured across each lamp (V_1 and V_2) then the total will equal V:

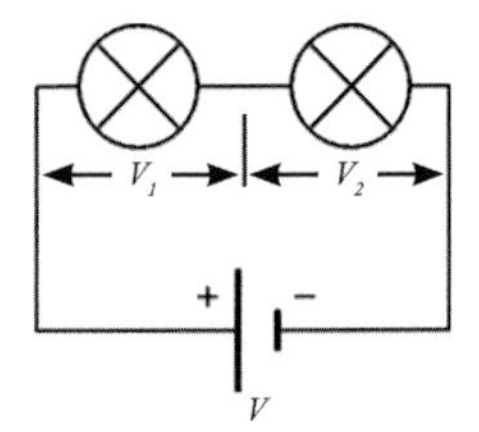

$$V = V_1 + V_2$$

Voltage in a parallel circuit

If the two lamps are instead placed in parallel (connected across each other) as now shown, it will be obvious that as they are both connected directly across the power supply then the voltage across each lamp will be the same and equal to the voltage of the power source.

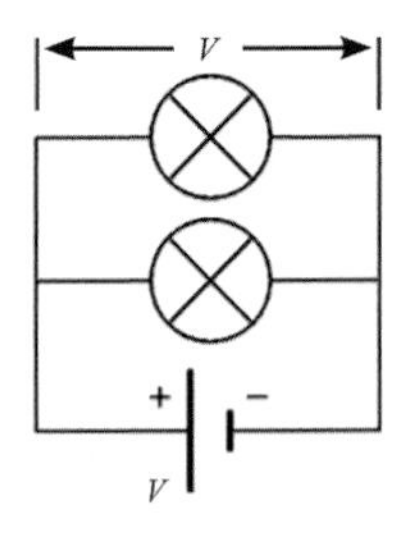

Always ensure that the positive and negative terminals of any power source are correctly connected, to avoid damage to the circuit or the power source itself.

No matter how many lamps are placed in parallel across the battery in the above diagram, the voltage across each of them will always be equal to the battery voltage. However, the battery will have to supply more current to light all of the lamps and will therefore discharge more quickly.

Power

When energy meets a resistance some of that energy will change into another form, usually into heat. This is what happens when current tries to flow through the wire inside a lamp. The *filament*, as it is called, is not a pure conductor and so presents a resistance to that current flow. In this case, the electrical energy is changed into heat and, as the filament warms up, is then seen as light.

In electronics it is quite common to deal with power in units of a few watts or less, such as milliwatt (mW) or microwatt (μW).

It is the change in electrical energy that makes lamps work; the amount of change, and hence emitted light, depending on the resistance of the filament.

The unit of power is the *watt*, and the symbol is W. When a current of 1 A flows through a resistance of 1 Ω, a power of 1 W will be developed.

Because power is always dissipated as current flows through an electronic component it is important that the component has the correct power rating. If rated too low, the component will get hot and eventually burn out. This should be especially remembered when designing a circuit, as too much total dissipated power could cause the whole circuit to overheat and fail, or worse, catch fire.

Calculating power dissipated

You only need two out of the three values of current, voltage, or resistance to be able to calculate the power in a circuit.

1. If you know the current and the voltage then:

   power = current x voltage

   $(P = I \times V)$

2. If you know the current and the resistance then:

   power = current x current x resistance

   or power = current^2 x resistance

   $(P = I^2 \times R)$

3. If you know only the voltage and the resistance then:

   power = (voltage x voltage) divided by resistance

   or power = voltage^2 divided by resistance

   $(P = V^2/R)$

In a series circuit, simply add the resistances together to find the total resistance for your calculation. In a parallel circuit, first work out the power in each component then add the answers together to find the total power.

Ohm's Law

Earlier in this chapter you learned that current is defined as the flow of electrons and that any opposition to this flow of electrons is called *resistance*.

The lower the resistance presented by a circuit then the greater the current flowing through that circuit. It therefore follows that the greater the resistance presented by a circuit then the lower the current flowing through a circuit.

Current passes more easily through some materials than others. Where it passes easily, the material is said to have a low resistance. Where it passes less easily, the material has a high resistance.

Conductors

Materials that have a very low resistance and allow current to flow easily are called *conductors*. Most metals are good conductors. Because it doesn't easily oxidize, gold is often used in connectors to ensure a good quality electrical coupling. Copper is extensively used for wiring, such as in a transformer, and for the connecting tracks on a printed circuit board (PCB).

Insulators

Materials that have a very high resistance to the flow of current are called *insulators*. Rubber and plastic are two examples of extremely good insulators as they allow almost no current at all to flow. This is why they are used to encase wiring or for electrical mains plugs, etc., where the greatest protection against exposure to electrical wiring and connectors is required.

Temperature and resistance

As stated earlier, when power is dissipated in a component or a circuit, this is often manifested as heat. With some materials, when they get warm or heat up then their resistance changes. This can be a problem when things go wrong with a component, sometimes leading to damage to other components as well in a circuit if the heating up causes the component's resistance to fall and hence the current flowing to rise.

Always make sure that the power rating of all components chosen for a particular circuit are high enough to avoid issues resulting from any heat generated, thus avoiding costly failures.

At worse, this can result in something called *thermal runaway*, whereby the current flowing continues to rise as the temperature rises to the point where the component "cooks itself" and possibly other components too!

However, this increase or decrease in resistance due to heat can be put to good use, such as protecting the output transistors in a power amplifier if they get too hot. These special components are covered in more detail in Chapter 3 when we look at Resistors.

The relationship between voltage, current, and resistance was investigated in 1827 by Georg Ohm. His findings became known as *Ohm's law*, which states that current flowing through a conductor that is kept at a uniform temperature is directly proportional to the applied voltage but inversely proportional to the resistance of that conductor.

Remember that Ohm's law is only valid on the assumption that temperature always remains constant.

Ohm's law can be expressed using the following formulae:

$V = IR$ or $R = V/I$ or $I = V/R$

You may find it easier to remember the three formulae for Ohm's law by using a simple triangle diagram, as follows:

1. To find the voltage, cover the V and you are left with I alongside R. Therefore, this means that $V = I$ multiplied by R

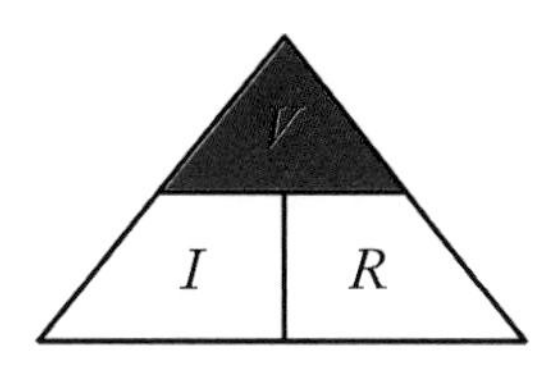

2. To find the current, cover the I and you are left with V above R. Therefore, this means that $I = V$ divided by R

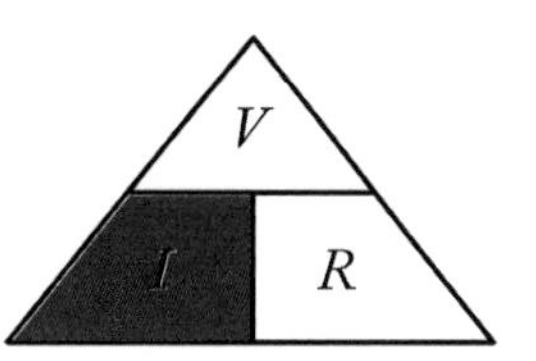

3. To find the resistance, cover the R and you are left with V above I. Therefore, this means that $R = V$ divided by I

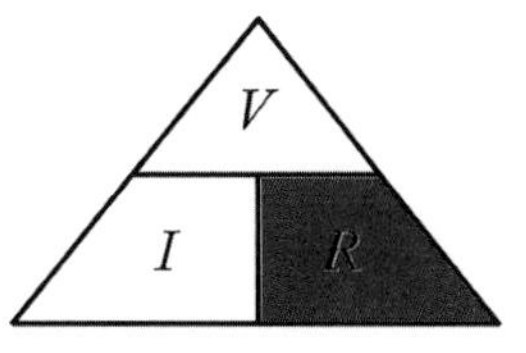

In electrical terms, resistance is defined as the opposition to the flow of electric current.

The unit of resistance is the *ohm* and the symbol is Ω. However, 1 Ω is quite small, so in electronic circuits it is common to refer to large values of resistance in terms of kilo- or mega-ohms.

A kilo-ohm is equal to one thousand ohms:

1 kΩ = 1000 Ω (or 1×10^3 Ω)

A mega-ohm is equal to one million ohms:

1 MΩ = 1 000 000 Ω (or 1×10^6 Ω)

Circuit Measurements

A formula like Ohm's law is extremely useful; if you know two of the values of voltage, current, and resistance you can quickly calculate what the third one should be.

You may need to do this when trying to establish why a particular circuit isn't working as it should. All may look correct, but how do you know that the value of a resistor, for example, is still what its color code says, and it hasn't changed for some reason? This can happen when a component has an insufficient power rating, or is placed close to another component that is getting too hot and consequently heating up everything around it.

A multimeter can be used to take readings of the voltage across a component and the current flowing through it. In the case of a resistor you can then use Ohm's law to work out the resistance and check if this matches the value of the resistor. If it does, then there may be an issue with the voltage or the current, and you will need to investigate the circuit further until the problem is located. The following circuit shows how to make simple measurements:

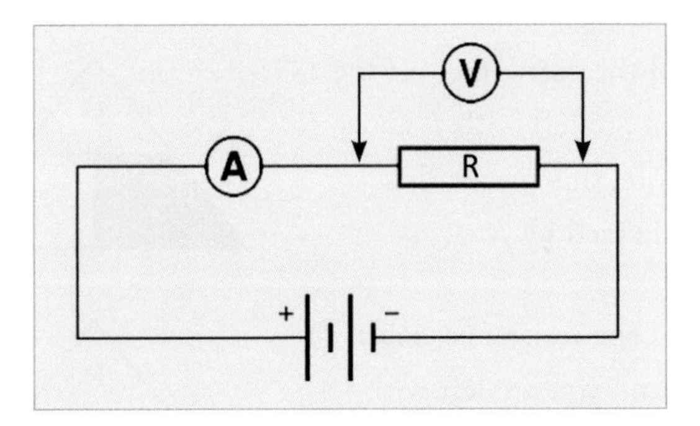

1. Set the multimeter to a suitable volts range and connect it across the resistor to measure the voltage

2. Set the multimeter to a suitable amps range and connect it in series with the resistor to measure the current

3. Using Ohm's law, calculate the value of the resistor – for example, if the voltage reading is 12 V and the current reading is 2 A, then resistance R = V/I = 12/2 = 6 Ω

Hot tip

If you are unsure what voltage or current range to set a multimeter to before taking a reading, simply start with the highest setting. You can always reduce it later for a more accurate reading.

To measure voltage, set the multimeter to a suitable voltage range and connect it across a component. Use the current setting and connect the multimeter in series with a component to measure the current.

Resistors in Series

So far we have only considered the simplest of circuits. Real electronic circuits normally consist of many components connected together to perform a particular function.

When there is more than one component in a circuit then the current will split and flow into these components in different amounts, though the current returning to the energy source will always be the same value as the current flowing from the energy source. The voltage across each component may also vary depending upon how they are connected together, as will the overall resistance or *load* that the circuit puts on the energy source.

Resistors connected in series

When resistors are connected end to end, as shown in the diagram below, they are said to be connected *in series*. Note the following:

When resistors are connected in series, the total resistance will be the sum of all the individual resistors.

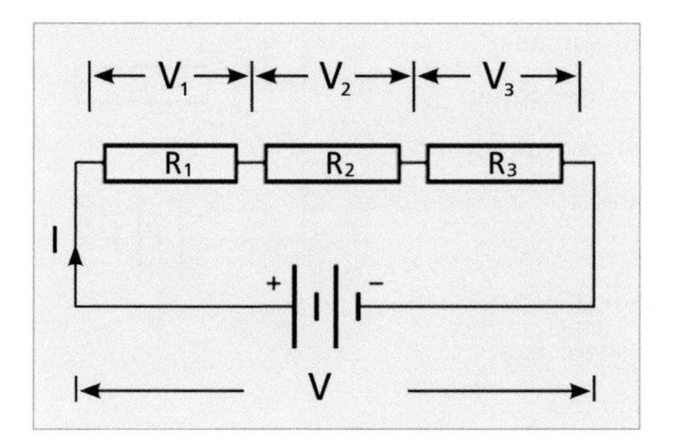

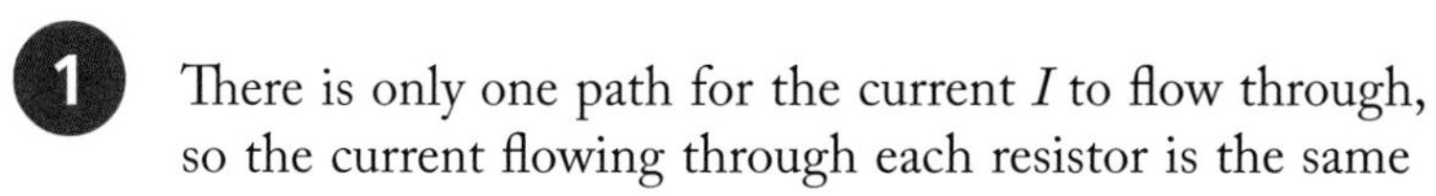

1. There is only one path for the current I to flow through, so the current flowing through each resistor is the same

2. The voltage $V = V_1 + V_2 + V_3$

3. The total resistance $R = R_1 + R_2 + R_3$

To recap, for resistors in series circuits:

- **The current is common to all resistors.**
- **Total voltage (supply voltage) equals the sum of the voltage across each resistor.**
- **Total resistance (circuit load) is the sum of all the resistors.**

Resistors in Parallel

Another way that components can be connected together is in parallel. The overall value of a number of components in parallel depends on the type of component; you don't just add their values together as is done for resistors in series.

Because current always takes the path of least resistance, most current flows through the resistor with the smallest value, whilst the least current flows through the resistor with the biggest value.

Resistors connected in parallel

If resistors are connected across each other and across the power source, as in the figure below, they are said to be connected *in parallel.* The current now has a number of paths to flow along, and splits into smaller quantities to flow through each resistor.

1. Because there is more than one path for the current I to flow along, it divides itself into smaller currents branching through each resistor (I_1, I_2 and I_3)

2. The voltage across each resistor is the same (V)

3. The current $I = I_1 + I_2 + I_3$

4. The total or equivalent resistance (R) created by the resistors in parallel is calculated using the formula:

$$\frac{1}{R} = \frac{1}{R_1} + \frac{1}{R_2} + \frac{1}{R_3}$$

As a general rule, the equivalent resistance of a parallel circuit is less than the value of the smallest resistance in the circuit.

To recap, for resistors in parallel circuits:

- **The voltage is common to all resistors.**
- **Total current (supply current) equals the sum of the current through each resistor.**
- **Equivalent resistance is the reciprocals of the individual resistance values.**

Series/Parallel Circuits

It is relatively easy to perform calculations when components are connected just in series or in parallel. However it becomes a little more complicated with real electronic circuits, because they normally consist of many components connected together in both series and parallel configuration. Take a look at the circuit below.

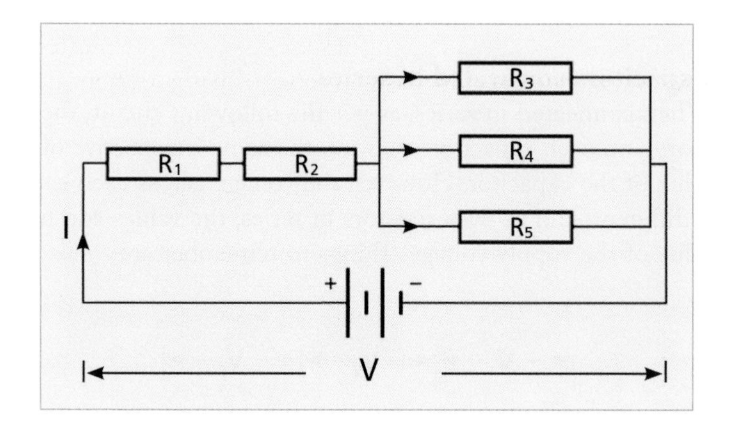

Replace the series and parallel parts of a circuit with their respective equivalent resistance to simplify calculating circuit values.

Now, there is a combination of two resistors in series connected to three resistors connected in parallel with each other. To work out the effective resistance of this configuration, do the following:

1. Initially, split it into two circuits consisting of R_3, R_4, and R_5 in parallel, and R_1 and R_2 in series
2. First, calculate the effective value of R_3, R_4, and R_5 using the formula for resistors in parallel from Step 4 on page 28
3. Now, simply add together the values of R_1, R_2, and the resulting value from the Step 2 calculation to give the effective value of the series/parallel circuit

Simplifying a circuit into smaller series and parallel circuits makes it easier to calculate the effective resistance in various parts of that circuit. For complex resistor circuits there are also other laws, such as Kirchhoff's laws, but they are beyond the scope of this book.

Calculations can be relatively simple with resistance, but not always so with other types of components, as you will learn next.

Knowing what the effective resistance should be in various parts of a circuit is very useful when looking for a fault.

Capacitors in Series

A capacitor is an electronic component that can store electrical energy; this energy being in the form of an electric field.

Capacitors are covered in detail in Chapter 3, but in its simplest form, a capacitor is just two conducting plates separated by an insulator called the *dielectric*, hence the symbol for a capacitor. Like resistors, capacitors can be connected in series or parallel.

Don't forget

The farad is the unit of capacitance, but it is a very large unit and in practice, capacitors in electronic circuits will have much smaller values measured in:

- microfarads (μF)
- nanofarads (nF)
- picofarads (pF)

Capacitors connected in series

When connected in series, as per the following circuit, then the charge on each capacitor is always the same irrespective of the value of the capacitor. However, the voltage across each capacitor is different, and as with resistors in series, the values add up to the value of the supply voltage. Things to remember are:

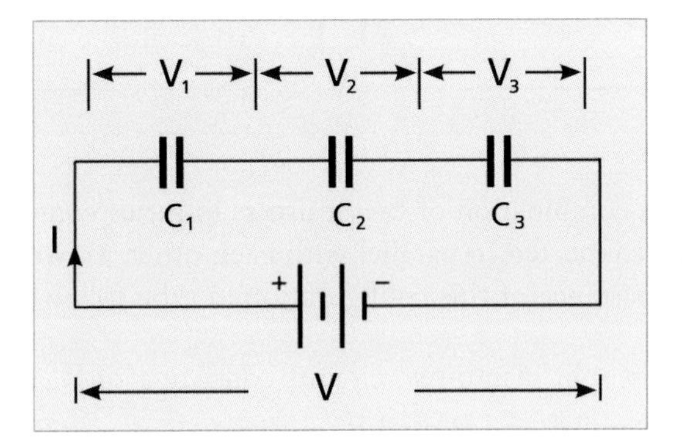

Don't forget

When capacitors are connected in series, the charge on each capacitor is the same.

1. The voltage $V = V_1 + V_2 + V_3$
2. The charge on each capacitor is the same as that drawn by the supply
3. The total capacitance is calculated using the formula:

$$\frac{1}{C} = \frac{1}{C_1} + \frac{1}{C_2} + \frac{1}{C_3}$$

Hot tip

The total capacitance of capacitors connected in series is calculated in the same way as for calculating the equivalent resistance of resistors connected in parallel.

Remember from Chapter 1 that the charge $Q = CV$, so having calculated the total capacitance in Step 3 then simply multiply the answer by the supply voltage to find the charge, which is measured in coulombs.

It follows that once you know the charge then it is a simple matter to calculate the voltage across each capacitor.

Capacitors in Parallel

A capacitor is charged when a voltage is applied to it. When capacitors are connected in parallel, the voltage applied to each capacitor will be the same, as shown in the diagram below, but the charge in each capacitor will be different and dependent on the value of each capacitor.

It follows, therefore, that for capacitors connected in parallel, the total charge provided by the power supply will be the sum of the individual charges in each capacitor. This is shown below.

When capacitors are connected in parallel, remember that the charge on each individual capacitor is directly proportional to the value of its capacitance.

1. Let Q_1, Q_2, and Q_3 represent the charge in capacitors C_1, C_2, and C_3 respectively

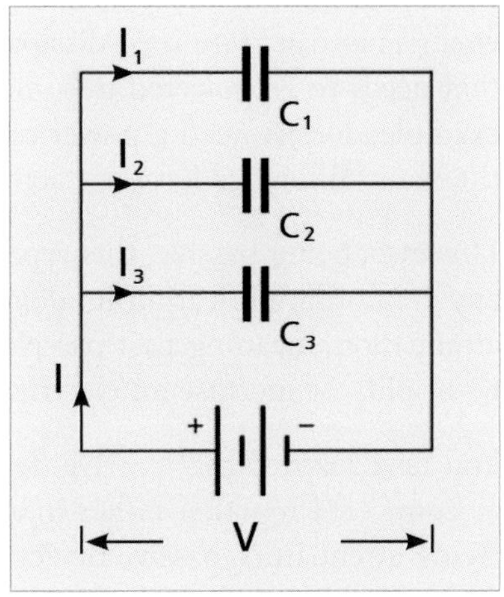
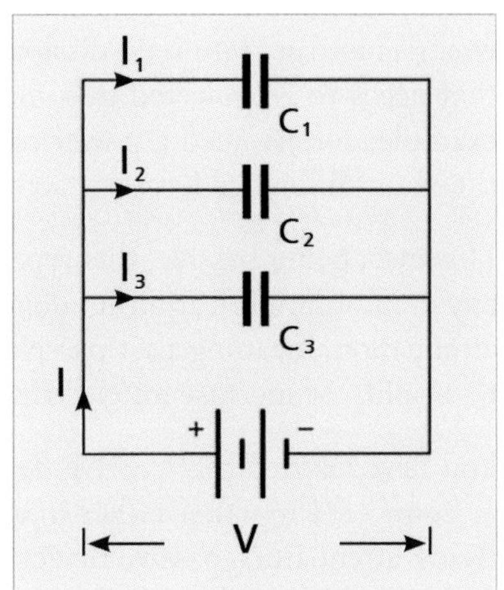

2. The voltage V is the same across all three capacitors

3. We know $Q = CV$ so CV will be equal to $C_1V + C_2V + C_3V$

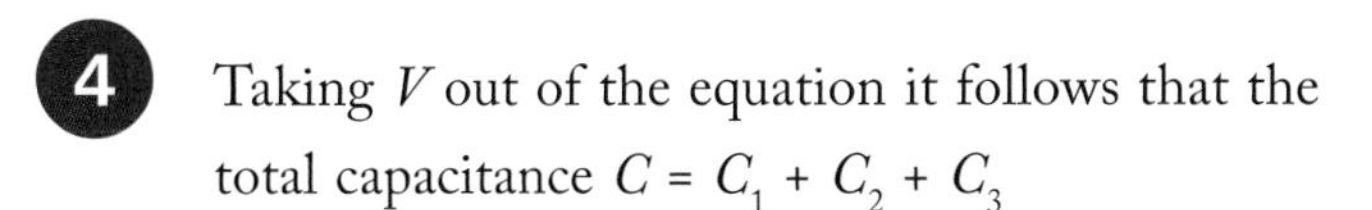

4. Taking V out of the equation it follows that the total capacitance $C = C_1 + C_2 + C_3$

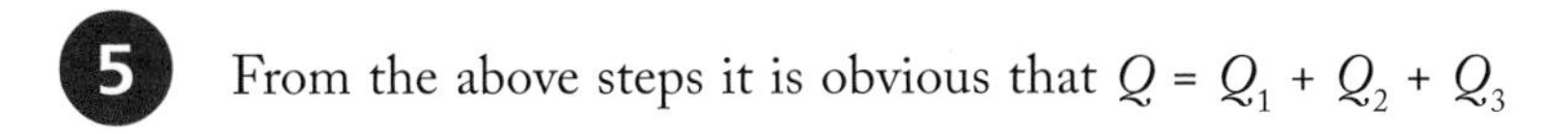

5. From the above steps it is obvious that $Q = Q_1 + Q_2 + Q_3$

For capacitors connected in parallel, the voltage across each capacitor is the same.

Note that the total capacitance of capacitors connected in parallel is calculated in the same way as for calculating the total resistance of resistors connected in series.

You should also be able to see that the charge on each capacitor, when connected in parallel, is directly proportional to the value of its capacitance.

That is about it for the basic calculations. You need to be familiar with these as they form the basis for understanding how electronic circuits and components work, or when trying to fault-find a circuit that isn't working properly.

Passive Components

The main difference between active and passive components is that active ones need to be powered in some way to make them work.

Components are grouped into two types. In this chapter you have learned about what are called *passive components*. The other type is *active components*, such as transistors and integrated circuits.

Passive devices are basically the main components used in electronics that together are required to build any electrical or electronic circuit. They include not only resistors and capacitors but also inductors and transformers, which will be covered later.

As the name suggests, passive components do not require any form of electrical power to operate. They do not in themselves generate energy, but can store it or dissipate it, unlike an active component that needs to be powered in some way for it to work. A resistor, for example, doesn't need a power source to provide resistance. This is the main difference between active and passive devices.

However, being passive, this type of device cannot generate any gain or amplification in a circuit. Instead, it can provide attenuation, meaning that passive devices in themselves are unable to amplify or increase an electrical signal.

You have learned that passive devices can be used individually or connected together either in a series or parallel combination. Being attenuators, passive devices can actually consume power within an electrical or electronic circuit, whereas active devices generate or provide power to a circuit.

Passive components can normally be wired into a circuit with no regard for polarity, whereas active components must always be connected with regards to positive and negative polarity.

Passive devices are usually bidirectional components, meaning that they can be connected either way around within a circuit with no regard for the positive and negative polarity of the power supply. Some specific types, such as electrolytic capacitors, do have a specific polarity marking, and care should always be taken to ensure that they are correctly wired into a circuit. The polarity of the voltage across them is determined by conventional current flow – in other words, from the positive to the negative terminal.

It is worth noting that component values of passive devices, such as resistance in ohms or capacitance in farads, are mostly positive in value (greater than 0) and rarely negative. There are exceptions to this, such as a thermistor, which has a negative coefficient, meaning that its resistance decreases with increasing temperature.

Active components and their function in a circuit will be discussed in Chapter 8 when you are introduced to the transistor.

3 Resistors and Capacitors

Learn all about the different types of resistors and capacitors, how they work, and what specific uses they have in various circuits. Tables are included for "at a glance" comparisons.

Types of Resistors

Resistors are passive components. This means they only consume power and cannot generate it.

Resistance limits the flow of electrons (and therefore current) through a circuit.

Metal oxide and metal film are now some of the most widely used forms of resistor. A wide range of values to a good degree of accuracy can be produced with both as well as with carbon film.

Now that the basics have been covered it's time to take a more detailed look at the various electronic components, the different types, and their specific uses. As before, we start with the resistor.

These fall into two major categories:

- **Fixed resistors** – the most widely used. There are many types of fixed resistor for use in different circumstances.
- **Variable resistors** – these consist of a fixed resistor element and a slider for tapping off a variable resistance.

Fixed resistors

There are quite a number of different types of fixed resistor; here are some of the more common:

- **Carbon film** – this resistor type is formed by depositing carbon onto a ceramic former; the resistance value being set by cutting a helix or spiral into the carbon film.

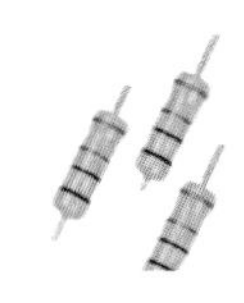

- **Metal film** – instead of a carbon film, this resistor type uses a metal film deposited on a ceramic rod.

- **Metal oxide film** – very similar to the metal film resistor, but uses a metal oxide film (such as tin oxide) deposited on a ceramic rod.

- **Wirewound** – generally used for high-power applications, these resistors are made using lengths of wire, called *resistance wire*, of which the exact resistance per length is known. The wire is wound round a former, and the resistor covered with a protective enamel coating.
- **Carbon composition** – formed by mixing carbon granules with an adhesive binder that is then made into a small rod. Once very common, but now seldom used as it is not very accurate.
- **Thin film** – uses thin film technology and is used for manufacturing billions of the tiny surface mount types of resistor (SMT) in use today.

Variable Resistors

This type of resistor lets you alter the resistance value needed in the circuit. The device itself has a maximum resistance value, and a wiper or slider device allows you to tap off a resistance between minimum and maximum.

Two types of potentiometers with different tracks are available: linear (lin) or logarithmic (log).

Also known as *potentiometers*, or *pots* for short, variable resistors are available in slider or rotary form. A typical volume control is a rotary type designed to be adjusted by the user, but there are also smaller types of variable resistors that, once set to a particular value, require no further adjustment. These are called *preset variables* and are normally small devices soldered directly onto a circuit board and set up at the manufacturing and testing stage.

An important point to remember is that variable resistors are available as two types – *linear* and *logarithmic* (or *log*). The term refers to the way the resistance changes as the wiper or slider moves across the track.

Linear

These are the most common. With a linear movement, the resistance changes in proportion to the amount that the slider moves along the track. For example, with a linear variable resistor of 1 kΩ then 500 Ω will be at the midpoint of the track.

Logarithmic

A logarithmic potentiometer means that the variable resistor has been manufactured in such a way that as the slider is adjusted, the resulting resistance changes in a logarithmic manner. A volume control is a good example of a log potentiometer; sound behaves in a logarithmic manner and hence requires log type control.

Because the response of the human ear to the loudness of sound is logarithmic then a logarithmic potentiometer must be used as volume control.

Preset

Variable resistors that can be adjusted by the user are usually easily accessible and quite large in comparison with a preset variable resistor. Preset variables are used to finely trim a voltage in an electronic circuit, and are usually adjusted by using a screwdriver. They are used to compensate for any variations in circuit values caused by component tolerances, and may be required in order to set the circuit to work within its required limits. Not accessible to the user, once set, a preset should not need altering.

...cont'd

There are various types of variable resistor, each with different properties and hence suitable for different applications or uses. Some have a composition very similar to fixed resistors. The following explains some of the more common types:

Cermet

The *cermet* variable resistor is one of the most popular in use today. The resistive element is made from **cer**amic and **met**al, hence the name. They have a low to medium adjustment life and so are widely used as trimmers.

Wirewound potentiometers are the most expensive type to produce. They are made by winding a coil of resistance wire on a semi-circular former. The surface of the wire must not be insulated so that the slider is able to make good electrical contact and avoid any "step effect" as it moves from one loop of the winding to the other.

Wirewound

Wirewound variable resistors are made using very fine resistance wire such as a nickel chrome alloy wound around a circular former. Because they offer a high level of linearity and close tolerance and are stable over a wide temperature range, wirewound variable resistors give a high level of performance and are a popular choice for audio applications.

Note that they do have a couple of main disadvantages though. One is that as the slider moves over the wires, the resistance changes in minute steps that may be a problem in some applications. Also, wirewound variable resistors are not suitable for some radio frequency applications, as the resistance wire forms a coil that could affect the performance of the circuit.

Carbon composition

This manufacturing process produces some of the least expensive types and so they are widely used in many areas, being a good all around general purpose variable resistor. Most linear and log potentiometers are of this type. Unfortunately, these variable resistors can become noisy after use due to wear and dirt building up on the track, though a squirt of switch cleaner often solves the problem.

Conductive plastic

Made using a conductive plastic ink-containing carbon resulting in a variable resistor with a high rotational life and a low noise output. They are often used for position sensors in servo-controlled machines.

Specialist Resistors

With fixed resistors it is very important that the resistance value is not affected by changes in the environment, such as a rise or fall in temperature. The same is true of variable or preset resistors; once set you do not want the resistance value to change.

However, there are some specialist resistors that have been manufactured in such a way that their resistance value does change with changes in their operating environment.

There are two types of thermistor: one where its resistance value increases as the temperature increases, and the other where its resistance value increases as the temperature decreases.

The thermistor

This is a device that has been designed to behave in the exact opposite way to that of the conventional resistor; that is, for its resistance to change in relation to variations in temperature. The name *thermistor* is derived from "**therm**al res**istor**".

There are two types of thermistor. One type is manufactured such that its resistance value increases as the temperature *increases* – this is called a *positive temperature coefficient* (*PTC*). With the other type, the opposite is true in that its resistance value increases as the temperature *decreases*, so it has a *negative temperature coefficient* (NTC). Believe it or not, the first NTC thermistor properties were discovered by Michael Faraday as long ago as 1833!

Because of how they work, resistance values for thermistors have to be stated in relation to a particular temperature range and coefficient – for example, NTC 20 kΩ (±5%) at -40 to +100°C.

The photoresistor

Also called a *light dependent resistor* or *LDR*, its resistance value is dependent on the amount of light it is exposed to. Typically, the resistance of an LDR decreases as the amount of light increases. A common application is for the automatic control of outdoor lights, allowing them to switch on as darkness approaches.

The most common type of LDR has a resistance that decreases with an increase in the light falling upon it.

Photoresistors come in many types, such as the inexpensive cadmium sulfide cells used in alarm devices, clock radios, camera light meters, solar street lamps, and nightlights, etc.

The humistor

If you have understood the above, you will have now guessed that a *humistor* is a special resistive device whose resistance varies depending on the humidity of the operating environment.

Electrostatics

As stated in Chapter 1, a capacitor is a device for storing energy in the form of an electrical charge. To understand this, an introduction to electrostatics is needed.

Don't forget

The term *electrostatic field* refers to the field of energy that exists between two objects of opposite polarity.

The electrostatic field

Electrostatics is the term governing the behavior and properties of an electric charge in the steady state. A charge can be positive or negative and, as in magnetism, opposite charges attract, whilst like charges repel. Remember that protons carry the positive charge and electrons carry the negative charge.

In magnetism, you have a magnetic field between magnetic poles. In electrostatics, the energy force between positive and negative charges is called an *electric* or *electrostatic field*. The energy contained in this field is simply called the *electrical charge*; the bigger the field then the greater the charge.

The term *electric field* refers to the area surrounding electrically charged particles.

The diagram opposite makes it easy to understand the electric field. Two metal plates are placed parallel to each other and a little distance apart. The red plate is charged with a positive potential and the blue plate with a negative potential. Negatively charged electrons will be pushed away from the negative blue plate towards the positive red plate. Positive charges are pushed away from the positive red plate towards the negative blue plate. The lines between the plates represent the electric force, and hence the direction of the electric field as it moves, by convention, from positive to negative.

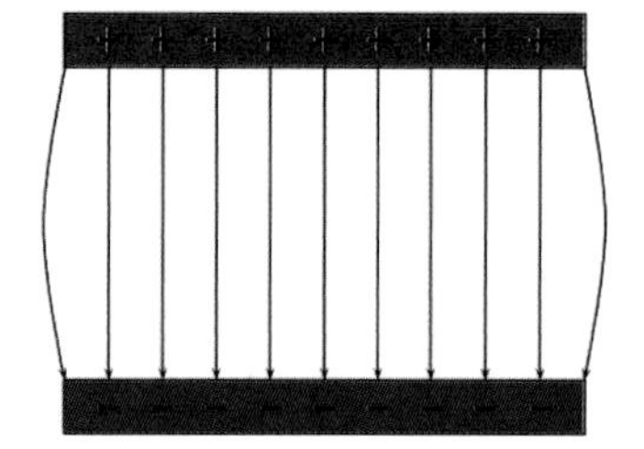

Guglielmo Marconi was an Italian inventor and electrical engineer best known for his extensive experiments into making long-distance radio transmission a reality. Jointly awarded the Nobel Prize in Physics for his radio work in 1909, he freely admitted in his acceptance speech that he didn't really understand how his invention worked!

This is basically what a capacitor is: metal plates that develop a charge between them when one is made positive and the other negative. For now, we can assume the plates are separated by air.

It is the above concept that Marconi used when he tried to send signals from one place to another without wires. He placed two large metal plates in the air a long distance apart and tried to fill the gap between them with an electric field. This did eventually work once he had a sensitive enough device to detect the received field. Later he found he didn't need the plates; the connecting wires were adequate to act as transmitting and receiving antennas!

Capacitance

Capacitance is defined as the property of storing electrical energy, and in Chapter 1 you learned that the unit of capacitance is the *farad* (*F*) and its symbol is C.

The simplest form of a capacitor is two parallel plates separated by an insulating material called the *dielectric*. In the diagram on the previous page the dielectric was assumed to be air, but many other materials can be suitable depending on the type of capacitor required and its use. In reality, a capacitor will consist of many plates, not just two, and can have a fixed or a variable value.

Here is the basic diagram of a simple two-plate capacitor. It shows the conductive plates separated by a dielectric, the thickness of which is labeled "d" for use in calculations.

Different materials have varying resistances to the forming of an electric field, but because the dielectric also acts as an insulator between the plates then its resistivity will always be very high.

Dielectric materials

The resistance of a material to the forming of an electric field is called its *relative permittivity*. The following table lists various common materials and typical values of relative permittivity.

Materials	Relative Permittivity
Air	1
Polythene	2.25
Polystyrene	2.4 – 2.7
Paper	1.4 – 3.5
Glass	3.7 – 11
Rubber	7
Water	80
Ceramics	6 – 1000

Capacitance is a measure of the amount of electrical charge that a capacitor is able to store between its plates for a given voltage.

Because the job of a capacitor is to store a charge, it is important to always remember to check that the capacitors in a circuit are fully discharged before working on them, especially those with a large capacitance value such as electrolytics.

Types of Capacitors

One of the most common uses of a capacitor is to let AC current pass whilst blocking DC current.

There are basically two types of capacitor – variable or fixed. Within each group, the capacitors come in many shapes and sizes and are generally designed for use in specific applications. For example, one type of fixed capacitor may be particularly suitable for use in audio circuits, whilst another type might be designed mainly for providing voltage smoothing in power supplies.

Because there are so many diverse types of capacitors it is not possible to describe all of them here, but sufficient to cover only the more commonly used ones.

Function

Here are some of the general functions capacitors are used for:

- Blocking DC but allowing an AC signal to pass through.
- Coupling an AC signal from one part of a circuit to another.
- Being used in a filter circuit to remove unwanted voltages.
- Smoothing or filtering ripple in a DC power supply.
- Making a simple timer when used with a suitable resistor.
- Storing a large charge such as for a camera flash.

Capacitor construction

There is a large number of types of capacitors, and the different ways in which they are manufactured is determined by factors such as dielectric, capacitance value, and working voltage required. The table below gives a comparison of these factors.

The term *working voltage* refers to the voltage that can be applied safely to a capacitor without the dielectric breaking down.

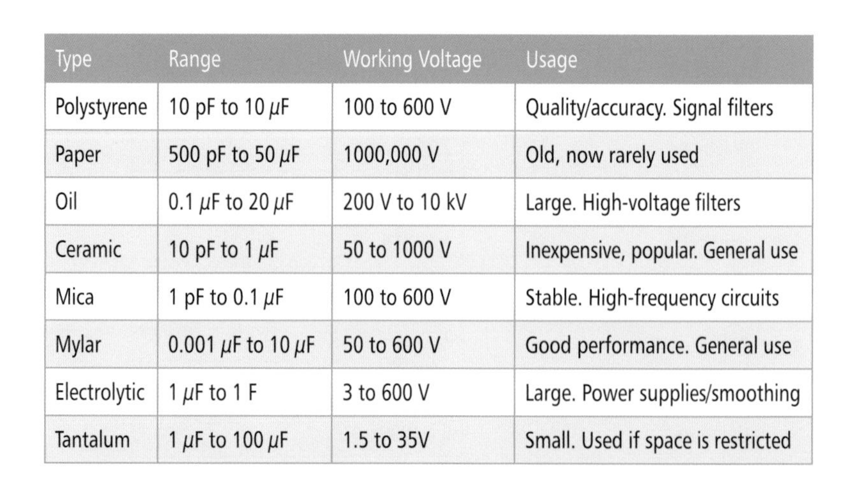

Type	Range	Working Voltage	Usage
Polystyrene	10 pF to 10 μF	100 to 600 V	Quality/accuracy. Signal filters
Paper	500 pF to 50 μF	1000,000 V	Old, now rarely used
Oil	0.1 μF to 20 μF	200 V to 10 kV	Large. High-voltage filters
Ceramic	10 pF to 1 μF	50 to 1000 V	Inexpensive, popular. General use
Mica	1 pF to 0.1 μF	100 to 600 V	Stable. High-frequency circuits
Mylar	0.001 μF to 10 μF	50 to 600 V	Good performance. General use
Electrolytic	1 μF to 1 F	3 to 600 V	Large. Power supplies/smoothing
Tantalum	1 μF to 100 μF	1.5 to 35V	Small. Used if space is restricted

Variable Capacitors

There are times when you need to be able to alter the value of a capacitor – for example, when tuning a circuit. There are three basic types of variable capacitor.

Air-spaced capacitor

As the name suggests, in this type of capacitor the dielectric is air, and you can see the metal plates, some of which are moveable. This capacitor was once very common, and generally used in the tuning circuit of a radio receiver, though this function has more or less been taken over by the significantly smaller (and cheaper) varicap. It is still used in transmitters for tuning the antenna.

Variable capacitors are often used to set the resonance frequency of a tunable circuit so are also sometimes called a *tuning capacitor*.

The air-spaced capacitor consists of a set of fixed plates and a set of movable plates mounted on a spindle. Maximum capacitance is when the plates are fully closed, and minimum capacitance is when they are fully open. The distance between the plates affects the maximum working voltage.

Mica trimmer capacitor

This type of capacitor, also known as a *compression trimmer*, is used where a very fine adjustment of capacitance value is required, such as accurately tuning radio frequency circuits or for adjusting the frequency of a crystal.

The dielectric is mica, and a small screw is used for adjusting the distance between the plates and hence altering the capacitance by a tiny amount.

Varicap diodes have no moving parts and are used mainly in radio frequency (RF) circuits to provide a capacitance that can be varied simply by changing the voltage on it.

Varicap diode

Also called a *varactor diode*, the varicap looks like a normal diode. The capacitance is dependent on the voltage across it, and the varicap is now widely used in the tuning of radios and TVs.

Variable Type	Range	Tolerance	Max. Working Voltage
Air-spaced	50 to 500 pF	±30%	500 V
Mica Trimmer	1 to 50 pF	±10%	100 V
Varicap	500 to 620 pF	±10%	10 V

Fixed Capacitors

Fixed capacitors are used in virtually all electronic circuits, with millions of different types manufactured every day. Each type has its own advantages and disadvantages that make it suitable for use in different applications. The following is just a small list of the common capacitor types that you are most likely to come across.

Most capacitors today have their value printed on them, but just like with resistors there is also a capacitor color code that can be used with some types.

Mica

Also known as *silver* or *silvered mica* capacitors. As the name implies, they use a mica dielectric which has silver electrodes plated directly on to the mica film to form the plates or electrodes of the capacitor. Close tolerance values are achievable with good temperature stability. Because of this it was once the capacitor of choice for many radio frequency circuits like oscillators and filters.

Ceramic

Ceramic capacitors have recently taken over from mica as they have a similar level of performance but are smaller and cheaper. Typically, their value can range from a few picofarads to around 0.1 μF.

Plastic film

Foil film – plastic separates two metal foil electrodes.

Metalized film – a very thin layer of metal deposited onto the plastic film.

If an electrolytic capacitor begins to get hot in a circuit it is probably operating above its working voltage or simply connected backwards.

Electrolytic

Most popular leaded type for large values, use thin films of aluminum foil and an electrolyte, polarized.

Tantalum

Like electrolytic capacitors but much smaller, use a film of oxide on tantalum, also polarized.

Fixed Type	Range	Tolerance	Max. Working Voltage
Mica	1 pF to 0.1 μF	±0.25% to ±5%	600 V
Ceramic	10 pF to 0.1 μF	±1% to ±20%	1000 V
Plastic Film	100 pF to 20 μF	±1% to ±20%	100 V / 250 V / 2000 V
Electrolytic	0.1 μF to 1 F	-20% to +80%	500 V
Tantalum	1 μF to 100 μF	±20%	35 V

4 Magnetic Principles

This chapter explains the relationship between magnetism and electricity, and how these two forces are made to work together. Learn how the electromagnetic effect is used to generate electricity, create motion, or operate a switch.

Properties of Magnetism

Like electricity, magnetism is one of those forces that you know exists despite the fact that you can't actually see it. What you *can* see is the *effect* of magnetic force in the same way that you don't actually see the electricity flowing through the filament of a light bulb, but simply the effect the electrical energy has on it; that is, the filament heats up and gives off light.

Early records show that magnetism was being used as a navigation aid in ancient China.

It was from noticing certain effects that they could not readily explain that early scientists realized that there were forces that demanded further investigation, and magnetic force was one of those things. The property of magnetism was actually first discovered centuries ago through lodestones. A *lodestone* is a naturally magnetized piece of a mineral called *magnetite*.

A piece of lodestone suspended on a string was observed to turn and line up in a certain direction. This was used as an early form of compass to aid in navigation – in fact, the name *lodestone* refers to a "leading stone". This in turn led to the investigation and discovery of the existence of magnetic fields, and particularly to the existence of the Earth's magnetic field.

A magnet is an object or material that produces a magnetic field. Sometimes the magnetic field is quite large and easily detected, but similarly, sometimes it is so small that it is very difficult to detect its existence.

Ferromagnetic material is the term given to a metal whose molecules move freely and so can be easily made to line up and so turn it into a magnet.

Many metals have a magnetic field. The effect is strongest in metals where the atoms are grouped together in a way that produces tiny individual magnets. When these miniature magnets are all correctly aligned, either naturally or through the influence of another magnet or electric current, the material itself becomes a magnet. Iron has excellent magnetic properties. Materials that behave in this way are called *ferromagnetic materials*.

Magnetism and electricity are very strongly linked. Whenever a current flows, a magnetic field is created. This effect can be put to good use – for example, in creating what is called an *electromagnet*. A coil is wound around a piece of metal that has virtually no magnetic field of its own. When a current is passed through the coil then a magnetic field is created. When the current is switched off, the magnetic field disappears. Very large electromagnets are often seen in use in scrap metal yards, such as for picking up a car body and dropping it into a crusher.

Magnetic Field

All magnets have a magnetic field that surround them and that is strong enough to have an influence on other materials. As the name suggests, a permanent magnet is surrounded by a permanent magnetic field and, as you learned on the previous page, is typically a piece of ferromagnetic material.

Everything has a magnetic field, including the Earth. The Earth's magnetic field is not actually symmetrical because the north magnetic pole is a few degrees offset from the North Pole and moves gradually over time.

However, not all ferromagnetic materials are permanent magnets. In some, the molecules only temporarily line up when the material is placed in the presence of a strong magnetic field to form a temporary magnet. When this magnetic field is removed, the material loses its magnetism, sometimes quickly and sometimes very slowly, depending on the metal.

The magnetic field is strongest at the ends of a magnet; these ends are known as the *poles*. Lines of magnetic force are said to flow out of one end, called the *north pole* (*N*) and into the other end, called the *south pole* (*S*). This is exactly how we envisage the Earth's magnetic field, which is why the terms north and south are used.

You can't see or touch this magnetic field but it is possible to observe its effect and hence get some idea of what it looks like and proof that it does exist. For this you take a bar magnet and place a piece of card over it. Iron filings are sprinkled onto the card. If the card is then gently tapped you begin to "see" the magnetic field of the magnet under the card as the iron filings move around slightly and join together in lines that represent that invisible magnetic field.

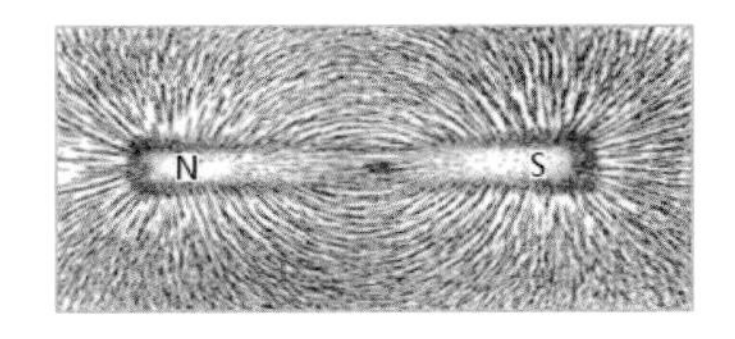

Opposite poles attract

If two bar magnets are placed with their ends facing each other then the effect of their magnetic force is very noticeable. When the north pole of one is placed facing the south pole of the other then the magnets are attracted toward each other.

Opposite poles attract, but like poles repel.

Like poles repel

When one of the magnets is turned around so that you now have two north poles or two south poles facing each other, the magnets try to push each other apart.

The above effect is used in a loudspeaker to produce sound.

Magnetomotive Force

Magnetic flux is the term that is given to the amount of magnetic field (or the number of lines of force) produced by a magnetic source.

Earlier it was stated that there is a very close relationship between magnetism and electricity; now it is time to examine it further. This relationship was first noted in 1819 when Danish physicist Hans Christian Oersted happened to spot that a compass on a nearby table kept deflecting while he was performing experiments with electrical currents in his laboratory.

Magnetic flux

If a coil wound around a ferromagnetic core has a current passed through it then a magnetic circuit is created. As the current flows through the coil, a magnetic flux is created that flows around the core. It is what is called *magnetomotive force* or *mmf* that causes this magnetic flux to exist in the magnetic circuit.

The amount of magnetic flux created depends on a number of interrelated things, such as the amount of magnetomotive force, the number of turns in the coil, the amount of current flowing around the circuit and, equally importantly, a specific property of the ferromagnetic core.

Permeability

The term *permeability* refers to how easily a material will magnetize in relation to the strength of the magnetic field acting upon it. Because a magnetic field will develop slightly quicker in a vacuum, then relative permeability is a more true indicator of the degree to which a magnetic material magnetizes in relation to the magnetic field acting upon it in a "normal" environment. The table below lists some common magnetic materials and a typical range of values of their relative permeability.

Beware

Not all metals will magnetize easily, whilst others can lose their magnetism quite slowly once the magnetizing force has been removed.

Magnetic Material	Relative Permeability
Ferrite (Nickel Zinc)	16 – 640
Cobalt	70 – 750
Nickel	100 – 600
Cast Iron	100 – 600
Steel	100 – 800
Rhometal	1000 – 5000
Permalloy	2500 – 25,000
Mumetal	20,000 – 100,000

Electromagnetic Induction

Many electrical or electronic devices rely on a principle called *electromagnetic induction* in order to function. The term refers to electricity being produced in a conductor by a changing magnetic field. To understand it better, it helps to go back in time.

This phenomenon was first observed by Michael Faraday in 1831 and, realizing that this was an important discovery, he carried out a number of experiments to determine exactly what was happening. He noticed that:

- When the magnetic flux linking a conductor changes then an electromotive force is induced in the conductor.
- The magnitude of the emf induced in the inductor is proportional to the rate of change of the flux linkage.

The above relates to one of Faraday's laws. In simple terms, it means that if a conductor is moved within a magnetic field then the force it experiences causes an electric current to flow through that conductor. Likewise, the faster the change in the magnetic field then the greater the emf produced.

The important thing to remember is that the electromotive force will only be produced in the conductor as long as the magnetic field is changing – therefore, no change, no current flow!

A conductor is the term given to a material whose electrons will move easily from one atom to another.

A typical conductor is a length of copper wire.

Other experiments

Michael Faraday was not the only one to have observed the phenomenon. Working independently, the American scientist Joseph Henry also discovered electromagnetic induction at about the same time as Faraday. In recognition, the *henry* (symbol H) is now the SI unit of inductance.

Many others quickly followed with further work in this field. In 1834, the German scientist Emil Lenz put forward what became known as *Lenz's law*. He stated that:

- The direction of an induced electromotive force is always such as to oppose the change producing it.

Apart from the production of electricity in a conductor, he also noted the existence of a mechanical force opposing the motion. As you will learn later, these laws are put to good use in a generator to produce electricity, and a motor to produce movement.

Electromotive Force

Although both Joseph Henry in America and Michael Faraday in England conducted experiments at about the same time, Faraday was the first to publish his findings and thus is the person credited with discovering electromagnetic induction.

What has so far been discussed about magnetism has been fairly general, and some of it you may already know. To fully understand the role that magnetism and its related forces play in electronics, it is time to take a slightly closer look at these invisible forces. Here is what you have learned in this chapter up to now:

When an object interacts with another object without physical contact it is said to be as a result of a *field*; something that produces a force. Electrical and electronic principles are based on two very important fields.

Electric fields

- Electrically charged objects are surrounded by an electric field.
- This field affects other charged objects.
- The charge on an object may be positive or negative.
- If objects have the same charge, they repel each other.
- If objects have opposite charges, they attract each other.

Magnetic fields

- Charged objects produce an electric field whether they are moving or stationary.
- When moving, an additional field is produced called a *magnetic field*.
- The magnetic field is a result of the spinning and orbiting of charged electrons around the material's atoms.
- These atoms are each equivalent to a tiny magnet with a north and a south pole.
- When the atoms are aligned in a random manner, their effect cancels out and the material is not magnetized.
- When the atoms are aligned in the same direction, their fields combine and the material is said to be magnetized.
- Lines of force called *magnetic flux* run from the north pole of a magnet to the south pole.
- As with electric fields, two like magnetic poles repel each other, while two unlike magnetic poles attract each other.

emf

Faraday's early experiments to show the existence of electromotive force (or emf) were crude, and initially he didn't actually use a magnet. However, it didn't take him long to realize there was a connection between electricity, magnetism, and movement and was soon using just a magnet and a simple coil to prove his findings.

His aim was to prove that if a conductor was connected as part of a closed circuit then an electric current would be observed flowing around the circuit as a result of the electromotive force produced in the conductor; this emf (or current) having been induced in the conductor as it passed through the lines of flux whilst moving across the magnetic field.

A galvanometer is an extremely sensitive meter capable of measuring very tiny amounts of current flowing through a circuit.

Faraday managed to do this by using a magnet and a coil of wire (as the conductor) connected to a *galvanometer* to complete the circuit. The zero of the galvanometer was at the center of the scale so the indicator could move to the left or the right of center to show in which direction the induced current was flowing.

When the magnet was pushed through the coil in one direction, the pointer on the galvanometer moved one way then back to the center point. When the magnet was pushed fully back through the coil in the opposite direction, then the pointer also moved in the opposite direction, then back to the center. If the magnet stopped moving at any time, then the pointer immediately returned to zero, proving the need for movement to induce an emf. Exactly the same results were obtained if the magnet was kept stationary and the coil moved instead.

Maxwell's right-hand grip rule

Eventually, the Scottish scientist James Maxwell introduced a rule for defining the direction of the circular magnetic field lines.

If the current-carrying conductor is gripped in the right hand with the thumb extended in the direction of the current (as shown), the wrapped fingers will show the direction of the circular magnetic field lines.

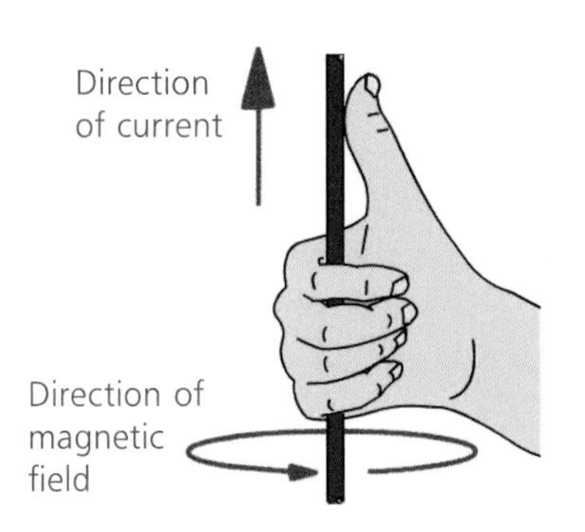

The right-hand grip rule

You can also use Fleming's right-hand rule to determine the direction of current produced by electromagnetic induction.

Inductors

There are two types of inductance: self (L) and mutual (M).

Having covered the link between electricity and magnetism, it is time to look at components that make use of this relationship. This is an important area, because if it wasn't for these components then facilities such as radio communication would not be possible.

Inductance

This relates to the property in an electrical circuit whereby a change in the current through the circuit produces an opposing electromotive force to that change in the electric current. The *henry* (H) is the unit of inductance, named after Joseph Henry.

- **Self inductance** (symbol L) – the emf is induced in the same circuit as that in which the current is changing.
- **Mutual inductance** (symbol M) – the emf is induced in a circuit due to a change of current in an adjacent circuit.

Inductor coils can be free-standing, wound around an iron or ferrite core, or wound around a large iron former.

Inductor (or coil)

An inductor is used where an element of inductance is needed in a circuit. It can take many forms, the simplest type being an *air-cored inductor* – just a coil of wire.

Where the coil is wound round an iron or ferrite core, the component is called an *iron-cored inductor*.

The above types of inductor are extensively used in radio frequency circuits.

Choke

When used in AC circuits, an inductor is often referred to as a *choke* because it "chokes" or limits the current flowing through it.

Be careful not to confuse a choke and a transformer as they can sometimes look similar. A choke should only have two terminals.

The number of turns, the way they are arranged, the cross-sectional area of the wire used, and the presence of an iron or ferrite core all affect the inductance of an inductor.

As with capacitors, the value of most inductors will be quite small so it is normal to see inductance expressed in millihenrys (mH) and microhenrys (μH).

Q factor

Inductances are also classified by something called a *quality* or *Q factor*. In simple terms, the higher the Q, the better the quality of the inductor. However, for some applications, such as when a broad bandwidth is required, a low Q inductor would be used.

The higher the Q factor of an inductor, the closer it approaches the behavior of an ideal inductor, and hence the better its efficiency.

Construction

As stated on the previous page, inductors are commonly classified by the type of core or former they are wound on. Air-cored inductors are generally wound on a plastic tube, having no electrical properties. Some are free-standing and have no former if the wire is sufficiently rigid.

You will come across air-cored inductors in radio applications, where they are used in tuning circuits and for matching the connection to the antenna. If used in a transmitter circuit, the inductor will be physically large to avoid any arcing between the turns due to the presence of high voltages. At higher frequencies such as VHF, UHF, and above, the inductors are usually quite small and often of the free-standing and self-supporting type.

Ferrite-cored inductors are so called because they are wound on an iron-based material called *ferrite*. The ferrite can take different forms, such as a ferrite rod aerial or a movable core known as a *slug*. The value of the inductor can be altered slightly by screwing the slug in or out of the former. This type is often used in resonant circuits so that the frequency can be accurately set.

In combination with capacitors, radio frequency (RF) inductors are used for making tuned circuits as used to tune radio and TV receivers. Inductors are also used in electronic filters to separate signals of different frequencies.

Ferrite is also available in a ring shape, and inductors wound on these cores are called *toroid inductors*. They are often used at low frequencies where comparatively large inductances are required. Because the core is a ring, and has no ends from which magnetic flux might escape, then all of the flux is contained in the core. This means that a toroid is less affected by nearby magnetic fields, and its own magnetic field has less effect on nearby components.

The laws of electromagnetic induction can also be put to a number of other uses, such as in transformers, electric generators, and motors. We will now look at each of these in turn.

Transformers

Remember that a transformer is an AC machine and cannot function on a DC supply.

A transformer is a device that does what it says – transforms an alternating voltage/current into another value of alternating voltage or current. The supply voltage must be AC for induction to occur because, as was discovered, electromotive force will only be produced as long as the magnetic field is changing. This obviously means that a transformer cannot work on a DC supply.

Whilst transformers come in all shapes and sizes – from large power transformers used in electricity generating plants to the small devices used in electronic circuits – the principle of operation is exactly the same in that the effect of mutual inductance is used to deliver a voltage at the output that can be higher, lower, or similar to the value of the voltage at the input.

In practice, a transformer can have more than one secondary winding, such as where two or more secondary voltages are required. One could even be for a high voltage and one for a low voltage, in which case the term step-up or step-down wouldn't strictly apply.

This diagram shows the basic principle. Two coils are wound around a ferromagnetic core; the input is called the *primary winding*, and the output the *secondary winding*. When an AC supply (V_1) is connected to the primary, the current that flows (I_1) causes a magnetic flux to flow around the core. This in turn induces an emf in the secondary winding to produce V_2 and I_2 to flow when a load is connected to complete the electrical circuit.

Magnetic flux
I_1
I_2
Primary winding
V_1
Secondary winding
V_2
Ferromagnetic core

Characteristics

- When the primary and the secondary windings have the same number of turns, then $V_1 = V_2$.
- When the secondary winding has more turns than the primary winding, then V_2 will be greater than V_1. This type of transformer is called a *step-up transformer*.
- When the primary winding has more turns than the secondary winding, then V_2 will be less than V_1. This type of transformer is called a *step-down transformer*.
- A transformer is rated by the volt-amperes it can safely transform before overheating occurs – for example, 40 VA.

The Electric Generator

An electric generator converts mechanical energy into electrical energy by means of electromagnetic induction – the phenomenon first noticed by Michael Faraday.

In theory it should be easy to generate electricity by moving a conductor through a magnetic field. In practice there are problems, such as how to move a wire at a constant speed through a fixed magnetic field. Because the most common means of propulsion is in a rotary form, the solution is to shape the wire or conductor into a loop. This can then be rotated inside a permanent magnetic field polarized north on one side and south on the other, as shown in the diagram below.

A copper slip-ring is connected to each end of the loop that, for practical purposes, consists of a coil of wire wound around a non-magnetic former. Carbon brushes are used to make contact with the slip-rings.

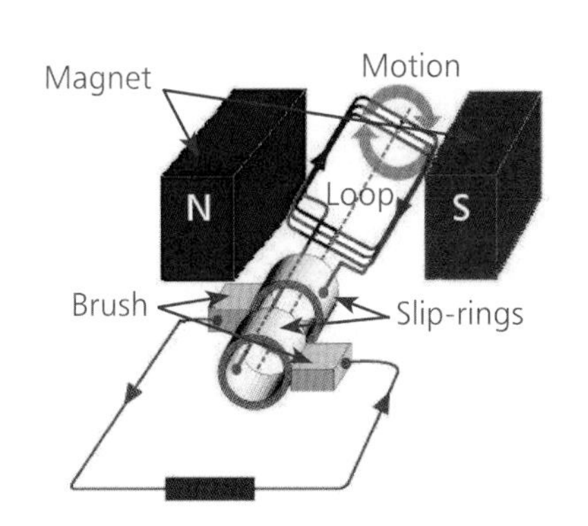

Eventually carbon brushes wear out and need replacing.

As per Faraday's laws, the emf (or value of the voltage) generated depends on the following three factors:

- The number of turns in the coil.
- The strength of the magnetic field.
- The speed at which the coil rotates.

Voltage generated

When the loop rotates within the magnetic field of the permanent magnet, the generated voltage begins to rise till it reaches a peak when the coil is in the vertical position. As the loop continues to rotate, the voltage then falls through zero to a peak in the opposite direction before beginning to rise back to zero as the loop completes a turn. This results in an alternating voltage (and current), as shown.

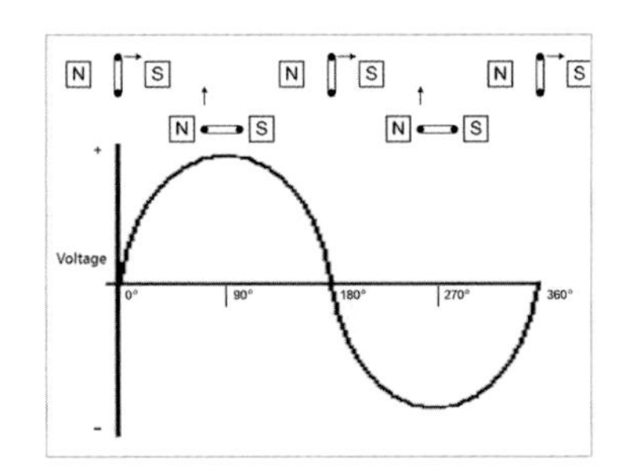

It is easier to generate AC than DC electricity. A DC generator doesn't really generate proper DC but a series of rising and falling half waves that then require some form of smoothing to keep the voltage at a constant level.

The Electric Motor

As with a generator, the carbon brushes in an electric motor will eventually wear out and need replacing.

Just like generators, electric motors are electrical machines. The difference is that with a generator, the input is in the form of mechanical energy, whilst the output is electrical energy. The reverse is true of a motor – electrical energy at the input results in mechanical energy at the output.

A simple motor

In an electric motor, the conductor is allowed to rotate freely in a uniform magnetic field. The diagram opposite shows a single loop conductor mounted between north and south permanent magnets.

Operation

When a DC voltage is applied to the loop, a magnetic field is generated that acts against the magnetic field of the permanent magnets. The resulting force causes the loop to start rotating, but as it does so, the force acting on it begins to lessen because the two magnetic fields are no longer in opposition.

To keep the loop moving when it has rotated through 180°, the current flow through the loop needs reversing so that the two magnetic fields are again in opposition. This current reversal is done by a *commutator*. Unlike the two slip-rings of a generator, the commutator is just a single ring that has been split down the middle to give two halves. Brushes are again used to maintain electrical contact with the split ring.

The function of the commutator is to reverse the electrical connection to the loop windings every half rotation.

Now, when a voltage is first applied, the current flows in one direction and the loop and commutator both rotate. When the commutator has gone through 180°, the electrical connections to the loop become reversed because of the split. The current now flows in the opposite direction, the magnetic opposition of the two fields is restored, and the loop continues to rotate – simple!

That is the basic principle. In reality, the process is made more efficient by splitting the commutator into not just two but many segments; each pair of segments having its own conducting loop. Also, the conductor will be a coil and not just a single loop.

The DC machine

You will have noticed that the simple generator produces AC electricity, whilst the simple motor is powered by DC electricity and hence is a direct current machine.

The rotating inner part of an electric motor is called the *armature*; the stationary outer part is called the *stator*.

In practice, this DC machine consists of two main parts:

- **Armature** – the rotating inner part. The armature consists of a steel or laminated iron core surrounded by the armature windings plus the commutator.
- **Stator** – This is the stationary outer construction containing the yoke (a steel ring) to which the magnetic poles surrounded by a field winding are attached.

Field winding

Using a field winding avoids the need for a permanent magnet. Instead, the required magnetic field is produced when current is passed through the winding to create an electromagnet.

In a DC machine, the field winding may typically be connected in series or in parallel with the armature.

Typical motors

Motors come in all shapes and sizes; from very large ones used in industrial applications to the very small ones as found in electric razors, model cars, drones, etc.

There are also specialist types such as stepper or synchronous motors. These are used in situations where very accurate movement or positioning is required, such as in control equipment, and CD/DVD and mechanical hard disk drives.

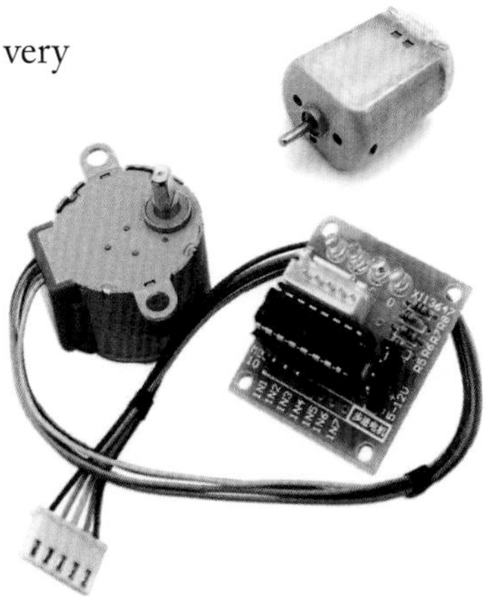

Additionally, there are also AC brushless motors, DC brushed motors, servo motors, linear motors and the newer high-efficiency direct drives that replace conventional servos.

Stepper motors are so called because the shaft doesn't actually continuously rotate but moves in precise "steps". They require some form of electronic circuit to provide pulses of electrical energy to produce the stepping motion, which can be in a clockwise or counterclockwise direction.

Relays

A relay is an electromechanical device because it is made up of a coil driven by an electric current (**electro**) and various moving parts (**mechanical**).

It is most commonly used as an electronic switch, especially where a low voltage needs to activate a much higher voltage. A good example is an outdoor security light where a low-voltage sensor circuit has to switch on an AC-powered high-wattage lamp when movement is detected.

Hot tip

When high voltage or current switching is required it is normal to use a low-power transistor circuit to activate a relay with heavy-duty contacts.

The operation of a relay is really quite simple. When a current flows through a conductor, the resulting magnetic field causes a set of contacts to change state by mechanical means.

Construction

The working parts of a basic relay are shown here. It consists of a coil wound around a ferromagnetic yoke, a set of contacts, and a pivoted armature. When a current is passed through the coil an electromagnet is created that pulls the armature toward it. This causes the armature to push on the movable contact so that the connection with the lower fixed contact is broken and connection is instead made with the upper fixed contact. When the current is removed, the magnetic effect is lost, the armature returns to its normal position, and the contacts swap back.

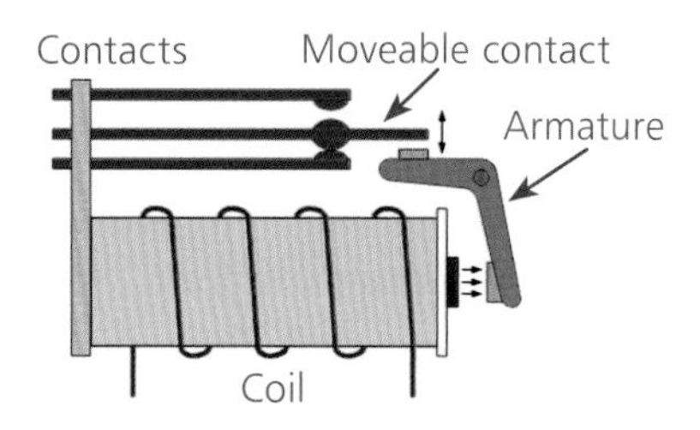

MBB is an abbreviation for *Make Before Break contacts* (also called *continuous contacts*). These contacts have an overlapping mechanism where the NO contacts close before the NC contacts open.

A relay may have just a few or many contacts, like PO3000 type relay once used in telephone exchanges (opposite). The contacts that are already touching are called *normally closed* (*NC*), and those that are open when the relay is off are called *normally open* (*NO*).

Reed relays

These are suitable for circuit board mounting, and consist of a tiny glass tube containing the contacts – sometimes multiple – placed inside an activating coil.

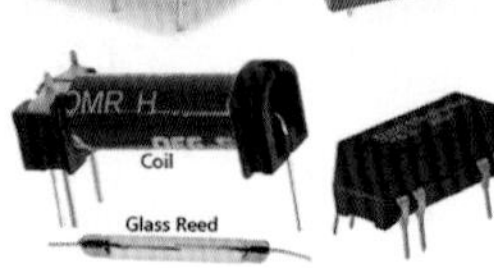

5 Single Phase AC Circuits

Learn how AC differs from DC, and of the phase relationship between current and voltage. Also learn new measurement terms and how waveforms can be observed with an oscilloscope.

AC (Alternating Current)

So far we have mainly considered voltage as being from a DC source, such as a battery where the voltage delivered is of a positive value even when the battery loses energy and needs recharging or replacing. The polarity is fixed; so, too, is the direction of the "direct current" as (by convention) it travels from the positive to the negative terminal.

Mains electricity is a term used in the UK and Canada, whereas US terms include wall power, domestic power, and grid power. They all refer to the alternating current (AC) electric power supply.

However, not all electronic circuits or devices use a battery as a power source. Many derive the necessary DC from an AC mains adapter that converts the high-voltage AC supply into a low-voltage DC supply to power the equipment. Such power supplies are covered later in Chapter 11, but first we need to look at alternating current in more detail.

Mains electricity supply

You will remember from the section on magnetism that alternating current electricity is generated by means of a coil rotating within a magnetic field. It is easier and cheaper to produce AC than DC, plus AC is far easier to distribute.

Domestic mains electricity in the UK is historically stated as a single phase 240 V AC supply (but see below) with a frequency of 50 Hz (Hertz) and is delivered to houses via the National Grid. Transformers at the power stations where the electricity is generated step up the voltage into many kilovolts so that it can be easily distributed along power lines with little energy loss. A sub-station then transforms the electricity to 240 V for domestic use.

The mains electricity supply differs around the world. In the USA, domestic appliances run on 110 V whilst European appliances are 220 V. Today's gadgets are "dual voltage", which means they work on both American and European supplies.

Outside of the UK the voltage and frequency of the electrical system varies by country or continent. From January 2003, the voltage used throughout Europe (including the UK) was harmonized at a nominal 230 V 50 Hz (formerly 240 V in the UK, 220 V in the rest of Europe).

Electric power is generated at either 50 or 60 Hz, and whilst a country will use one or the other it is still possible to find a mixture of 50 Hz and 60 Hz supplies in use in some countries. Years ago, some parts of the UK even had a DC mains supply!

The alternator

Unlike a DC voltage that remains steady, the value of an AC voltage is constantly varying. The polarity and value of the generated voltage is dependent on the orientation of the coil in relation to the north and south poles of the surrounding magnetic field. For one completed revolution of the generator, the resulting emf will have alternated between a maximum positive and a maximum negative value.

A generator is often called an alternator because it produces an emf that alternates between a positive and a negative value.

Because the value of the emf alternates between positive and negative, the generator is also called an *alternator*. You may know, for example, that on a motor car an alternator provides a charging voltage for the battery whenever the engine is running.

Waveforms

The term *waveform* is the name given to the shape of the resulting image when the value of the generated emf is plotted against time in the form of a graph.

The waveform on page 53 shows it starting at zero then alternating from a positive value, back through zero to a negative value, and then back to zero again. Only the first cycle of the waveform is shown; in practice, the process will keep repeating for each full rotation of the loop. This type of waveform, where it is constantly changing direction, is called an *alternating waveform*.

However, there are some waveforms that are said to flow in one direction only. In other words, although the value may change with time, it never actually crosses the time axis to go negative. This type of waveform is called a *unidirectional waveform*.

There are also three other waveforms often seen in electronics:

- Square wave
- Triangular wave
- Sawtooth wave

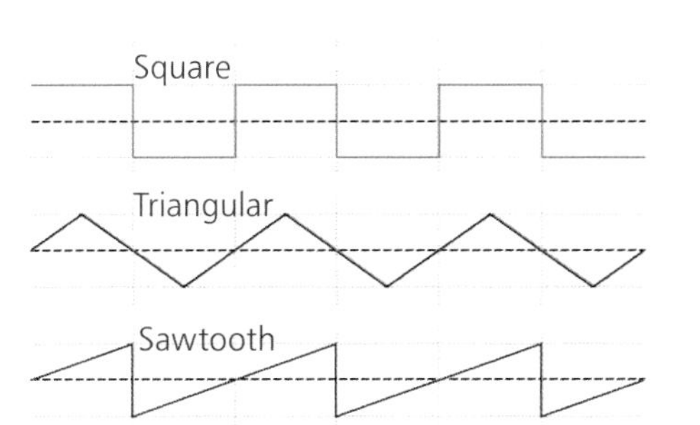

It will be obvious that the names relate to the shape of the waveforms. A square wave is most common in digital electronics.

There are many different types of waveform, not just the three mentioned on this page.

...cont'd

The sine wave

Because the waveform produced by a generator is sinusoidal in shape it is called a *sine wave*, and one complete 360° rotation is called a *cycle*. The *frequency* is the number of cycles completed in one second, and is measured in hertz (Hz).

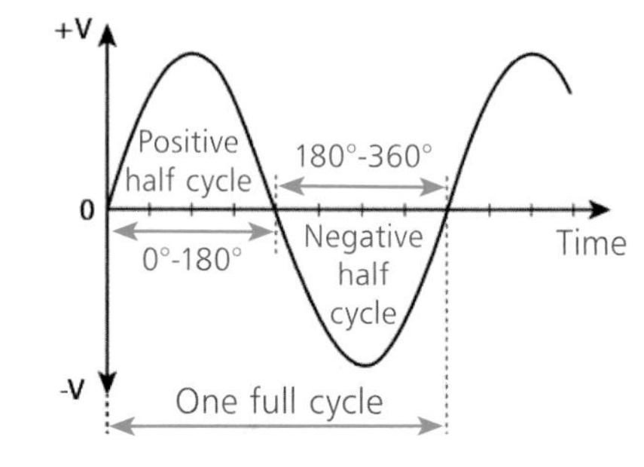

AC values

The value of alternating voltage or current is always changing and so can be expressed in a number of ways, such as peak-to-peak, peak, rms, etc. These ways of measuring AC are explained below.

The value of an alternating voltage or current is not as easy to measure as with DC because it is constantly changing relative to time.

Value	Description	Example / Formula
Peak-to-peak	The difference between the positive peak value and the negative peak value	Peak value; Peak to peak value; Peak value
Peak	The maximum value reached in either a positive or a negative half cycle	
Root mean square (rms)	The value of alternating current that will produce the same heating effect as an equivalent direct current	For a sine wave, rms value = 0.707 x maximum value
Average	The average of all the instantaneous measurements in one positive half cycle	For a sine wave, average value = 0.637 x maximum value
Instantaneous	The value of voltage or current at one particular time instant	If measured at the instant that the cycle polarity is changing then this value would be zero

AC values are commonly expressed as rms (Root Mean Square).

AC Terminology

Measuring a DC voltage is easy because it is a steady value, but measuring an AC voltage is a little more difficult as it is constantly varying. From the previous page it will be obvious that the different ways used to express an AC voltage are actually all referring to which part of the waveform we mean.

As well as the values explained earlier, here are some other AC terms you need to be more familiar with. For reference, use the sine wave diagram on the opposite page.

Cycle

This is the term given to one complete sine wave that, as stated, takes 360°. This doesn't have to start and finish at zero – a cycle is from any point on a waveform to the same point 360° later on the following waveform.

Periodic time

The duration of one complete cycle is called the *periodic time*. It is expressed in seconds or parts thereof.

Frequency

Frequency is the term used to define the number of cycles that occur in one second. Logically, frequency was initially expressed in *cycles per second* but this was later changed to the SI unit *hertz* (*Hz*) in recognition of Heinrich Hertz, the first person to provide conclusive proof of the existence of electromagnetic waves.

You may often come across reference to high frequency or low frequency. What this means is that there are more cycles in one second for a high frequency and fewer cycles per second for a low frequency. Periodic time and frequency are related, as follows:

periodic time = 1/frequency   frequency = 1/periodic time

Wavelength

Wavelength is the length of one cycle and is measured in meters. The term is generally used when working with radio or sound waves, though you may be more familiar with these waves being referred to by their frequency.

Radio waves travel at the same speed as light because they are both examples of electromagnetic radiation; the only difference being that they have very different wavelengths. The higher the frequency, the shorter the wavelength. Microwaves are even measured in centimeters (cm).

By convention, radio stations are often known by the wavelength they transmit on for the Long and Medium waveband, but on Short wave, VHF, and higher they are known by their transmission frequency.

Series AC Circuits

AC doesn't just mean *alternating current*; remember that the voltage is alternating as well. The relationship between the waveforms for current and voltage depends on the level of resistance, capacitance, and inductance in an AC circuit.

The industrial electrical supply is usually three-phase, whilst the domestic supply is single-phase.

Phase

It is now time to just briefly look at this relationship and to introduce a new term, namely *phase*, to identify whether the waveforms are in or out of step with each other.

- In phase (or step) is when the two waveforms are synchronized and their peaks and zero points match up at the same points in time.
- Out of phase (or step) is when their peaks and zero points do not match up at the same points in time, and so the waveforms are out of synchronization.

Unlike DC circuits, when inductors or capacitors are involved in an AC circuit, the voltage and current do not peak or cross the zero point at the same time, but are out of phase. The period difference between the peaks is expressed in degrees and is called the *phase difference*. This difference can also be shown graphically by means of a phasor diagram.

Purely resistive

Where an AC circuit is purely resistive, the voltage and the current are both in phase.

The diagrams opposite show the simplest purely resistive circuit, the AC waveform, and the phasor diagram for this circuit. Note that the voltage and current are synchronized and peak at the same points.

The phasor diagram is used to show that the voltage and current progress along the same time line, and are therefore in phase with each other.

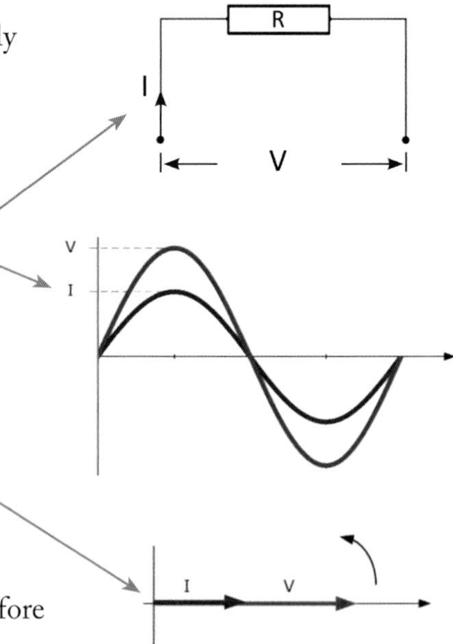

Use a phasor diagram to show the phase relationships between two sine waves of the same frequency.

Purely capacitive

Where an AC circuit is purely capacitive, then the current leads the voltage by 90°. From the waveform you can see that when the current is at a maximum, the voltage lags behind it and is still at zero. The opposition to the flow of alternating current is called the *capacitive reactance* (X_C).

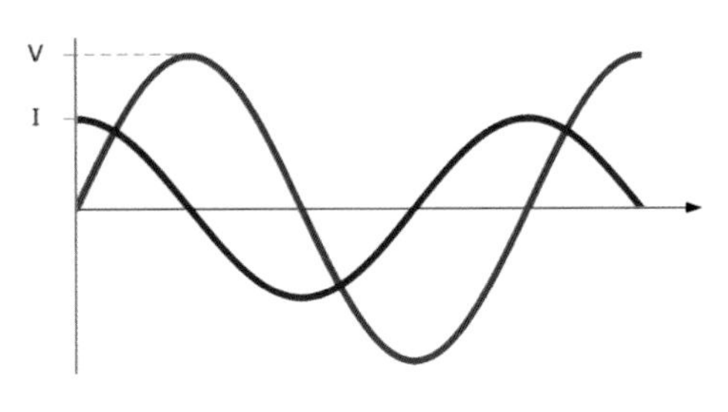

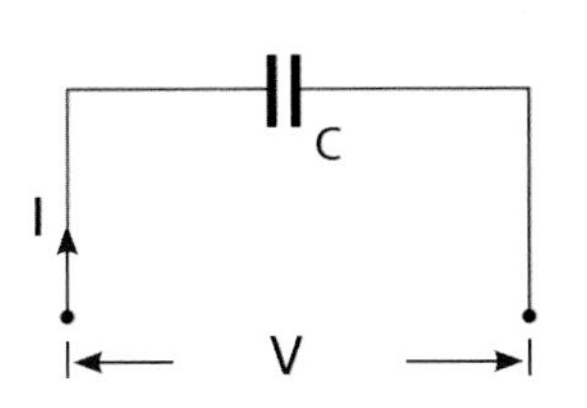

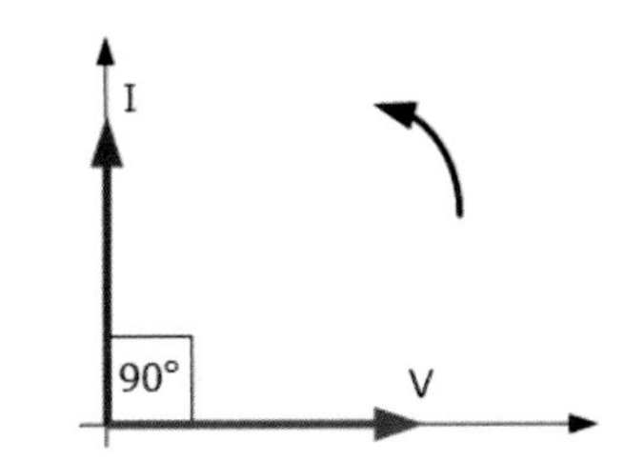

Don't forget

For AC circuits, capacitive reactance is represented by the symbol X_C, and inductive reactance by the symbol X_L.

Purely inductive

Where an AC circuit is purely inductive, then the current now lags the voltage by 90°. This time, the waveform shows that when the voltage is at a maximum, the current lags behind it and is still at zero. Now, the opposition to the flow of alternating current is called the *inductive reactance* (X_L).

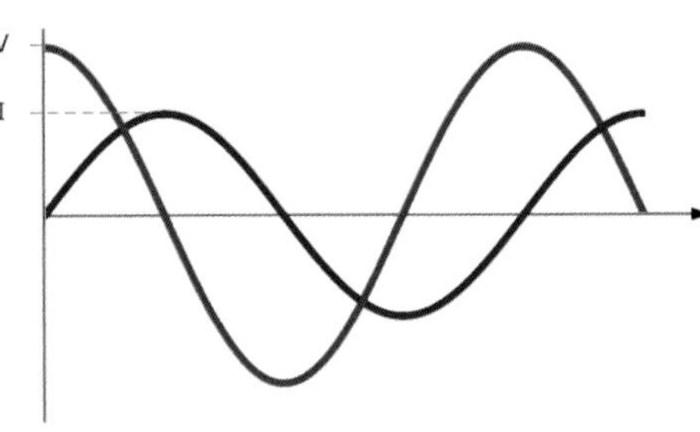

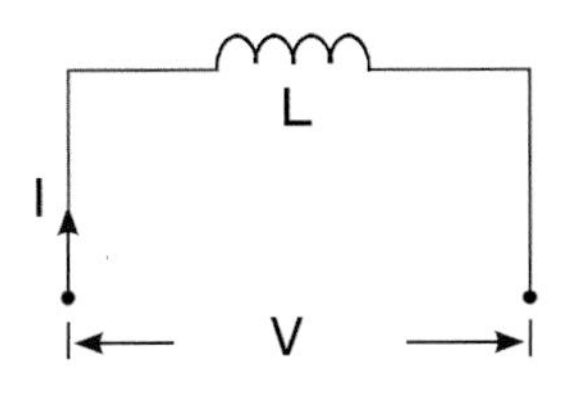

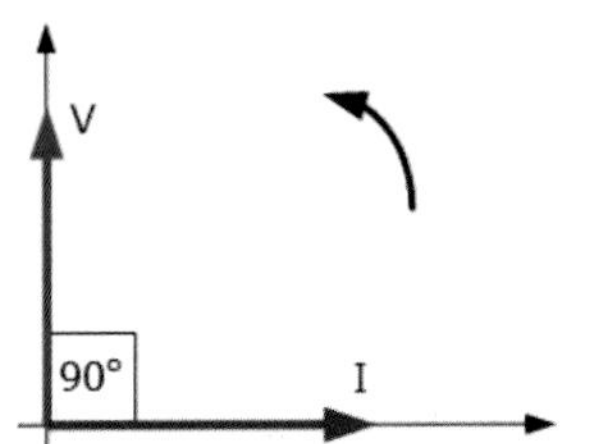

Don't forget

The current leads the voltage in a purely capacitive circuit but lags behind the voltage in a purely inductive circuit.

An AC circuit can also contain a mix of resistance, capacitance, and inductance, all of which will have an effect on the waveform.

R, L and C in Series

It should be remembered that in theory we tend to assume pure capacitance or inductance, but in the real world things are not so straightforward, especially when it involves AC.

Manufacturing processes make it difficult to have components that are purely inductive or purely capacitive, so stray values have to be taken into account when performing accurate circuit calculations.

In a capacitor, for instance, there could be an element of induction from the rolled construction of coiled plates, and presence of leakage resistance of the dielectric. Similarly, an inductor will have a resistive element due to the resistance of the wire forming the coil, plus some capacitance that can exist between adjacent turns.

Extensive theory is beyond the scope of this book but it is worth very briefly looking at the effect of these "stray" values of resistance, inductance, and capacitance as it introduces another common term.

An inductor is basically a coil of wire. If you tried to measure its resistance with a meter, the reading would read zero because the wire would be just like a short circuit, and the resistance far too minute to measure.

But in an AC circuit you learned on page 63 that there is reactance – a definite resistance to the flow of alternating current. Like everything else with AC this isn't so easy to measure, but it can be calculated when you know the inductance and frequency. Because it is a resistance, reactance is also measured in ohms, whether it is inductive reactance or capacitive reactance.

Impedance

When you connect a resistor and an inductor in series in an AC circuit you actually have two resistances in series: the pure resistance, and the reactance of the inductor. Unfortunately you cannot just add the two values together, as this is AC and there will be a phase difference between the voltage across, and the current through, the inductor. It has to be calculated using a vector diagram and Pythagoras's theorem.

Because reactance and impedance are all a form of resistance, the unit for both is also the ohm.

This combination of inductive reactance and resistance is called the *impedance* of the AC series circuit. The symbol for impedance is Z and again, because this is a resistance, the unit is the ohm.

The same is also true when a resistor and a capacitor are connected in an AC series circuit. This time, there is a combination of capacitive reactance and resistance to give circuit impedance. No matter how complex the combination of R, L, and C in AC series or parallel circuits, impedance is always the effective resistance.

Power in an AC Circuit

If it is a purely resistive AC or DC circuit then power can be calculated in the standard way using $P = IV$ (see page 13). However, it becomes more difficult to calculate power when an AC circuit also contains inductance or capacitance, as the current and voltage are always out of phase.

Unlike DC circuits, calculations for AC circuit values are more complex due to the constantly changing nature of the energy source.

Power dissipated in L and C circuits

Theory tells us that in an AC circuit of pure impedance or pure capacitance, no power will be dissipated because voltage and current are always 90° opposed. All of the power delivered to the circuit in one part of the waveform cycle is returned to the power source in the next part of the cycle.

Thus, in a pure L or C circuit, the average power dissipated must be zero, as the positive and negative cycles cancel each other out. Purely reactive components therefore should not dissipate any power, because all of the energy fed to the component and stored in the form of an electrostatic or magnetic field during one part of the waveform cycle promptly returns to the source in the other part of the cycle.

Apparent power

We know that in reality, AC circuits contain impure capacitance or impure inductance to complicate things. An appropriate meter could be used to measure the value of the supply voltage and how much current is being drawn from the supply.

Multiplying these two values together using $P = IV$ will give the *apparent power*, not the true power as, this being an AC circuit, the current is determined not by the circuit resistance (as with DC) but by the impedance of the circuit.

Because it is the product of an rms current and an rms voltage, apparent power values are not quoted in watts but in volt amps.

By ignoring any reactive elements in an AC circuit, the true value of power dissipated will always be less than the calculated apparent power.

True power

It is possible to calculate the *true power* from the voltage developed across, and current flowing through, just the resistive element of the AC circuit because this is where the voltage and current are in phase. Multiplying these two values together gives a more accurate power figure for the circuit, but because any reactive elements in the circuit have been ignored then the true value of power dissipated will always be less than apparent power.

AC Circuit Measurements

As you have seen, AC circuit measurements can be a multitude of things such as peak, peak to peak, amplitude, time, frequency, etc. Test equipment is covered in detail in Chapter 12, but it is appropriate to briefly introduce some of it now.

Some instruments can only measure one or possibly two AC values, whilst there are others that can measure most or possibly all of these AC parameters.

Don't forget

A multimeter is a very versatile instrument featuring a wide test range for voltage (both AC and DC), current, and resistance. Some also have the facility for testing diodes and even NPN or PNP transistors.

See pages 134-137 for more detailed information about multimeters.

Multimeter

The very useful multimeter can measure DC voltage, current, resistance and, if a digital instrument, possibly capacitance. For AC, an analog multimeter can usually only measure AC voltage.

- Analog
- Digital

Frequency counters let you measure frequency over a wide range, though you cannot actually see the waveform itself. See page 142 for more information.

Frequency counter

A different instrument is required to measure frequency; this is called a *frequency counter*. There are many types depending upon the frequency range the instrument is to be used for measuring. They all have a specific frequency range, though on some you select the range manually, whilst others can switch automatically.

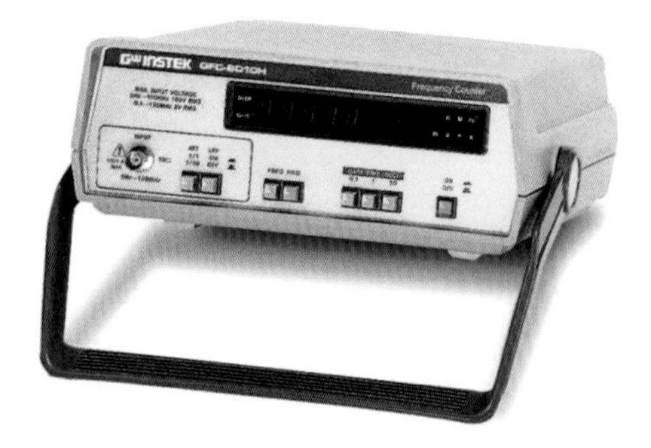

Oscilloscope

Although you can measure frequency with a frequency counter, of more use would be an instrument that actually lets you see the waveform. This is what an oscilloscope does.

Often simply referred to as a *scope*, the oscilloscope is a very versatile test instrument that can be used for a variety of applications. As you can see from the images, it has a display screen where you can look at waveforms and analyze them as required. For instance, you can see the shape of a waveform, if it is symmetrical, or whether it is affected by any distortion or noise.

An oscilloscope is very useful for observing if a waveform is clean and symmetrical or if it is distorted in any way. See pages 138-139 for more information.

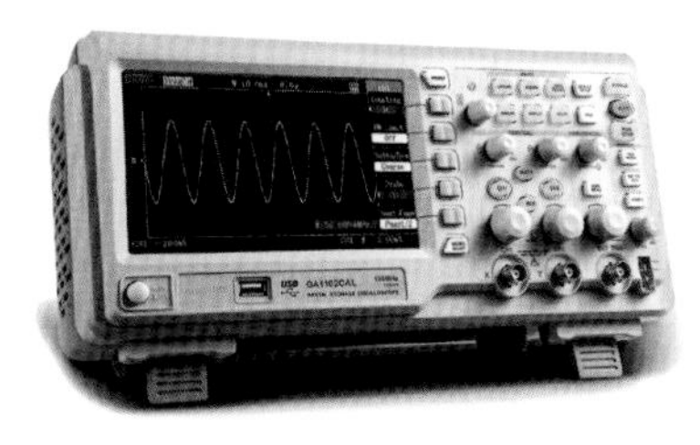

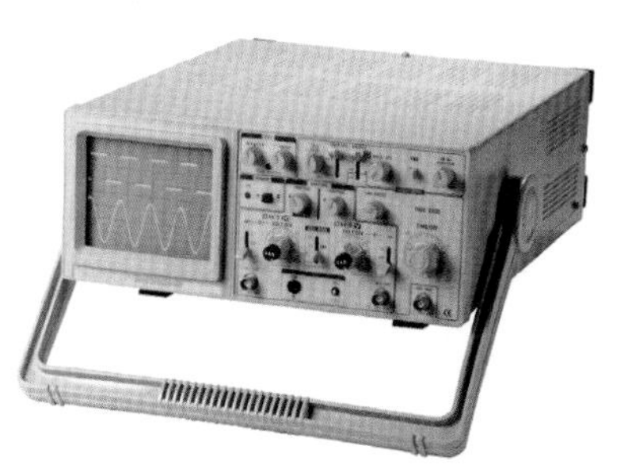

Traditionally, an oscilloscope was quite a large and heavy instrument that used a cathode ray tube (CRT) for the display. Although basic, a 1 cm grid on the screen together with various front panel switch and control settings allow the user to make relatively accurate voltage level and frequency measurements. There are still many of this type in use today.

With advances in digital technology, oscilloscopes have now become digital instruments that are considerably smaller, lighter, and more accurate than the CRT models. A digital oscilloscope can not only display the waveform but also display other information such as voltage level, frequency, and time, thus saving the user from having to calculate these from the grid.

Small low-cost digital oscilloscopes are now readily available in kit form from various sources on the internet.

Ringing

An advantage of being able to see the waveform is that you can tell at a glance how clean it is. This is very useful when fault-finding, as distortion or noise would not show up on other instruments but might be causing a malfunction. Ringing is one such irregularity and shows up as small oscillations as a fast waveform rapidly changes state, such as a square wave switching from low to high.

Rectifier Circuits

Chapter 11 covers power supplies in detail, but as a power supply is normally used to provide a constant DC voltage from an AC voltage source then we need to have a brief look at AC rectification at this point. This will help in describing how an oscilloscope can be used to observe the various waveforms.

A rectifier circuit is where an AC power source is converted into DC, usually at a much lower voltage. Because the AC supply will normally be the electrical mains supply, a transformer is first used to reduce the supply from the higher mains voltage to a lower voltage that can then be rectified.

Half wave rectification is the simplest way of converting AC to DC but is very inefficient.

Because the AC waveform is converted to a steady DC voltage, a rectifier circuit is ideal for investigating with an oscilloscope. You can observe the input sinusoidal waveform, the rectification action of a diode, and the resulting DC output. The oscilloscope also allows you to see if the output is steady or if there is any ripple that hasn't been successfully smoothed out.

Half wave rectification

This circuit shows a transformer with its input winding connected to an AC power source. A diode (D) is connected to the output winding, and the circuit is completed by load resistor R. If an oscilloscope is connected to any point on the input side of the transformer, the scope trace will show the full AC waveform cycle shown here.

However, because a diode only conducts on positive half cycles, if the oscilloscope is now connected to the output of the diode (where D and R are connected together), the waveform will look like this, proving that the diode does indeed only conduct when the cycle is positive and there is no output when negative. The scope gives visual proof of half wave rectification.

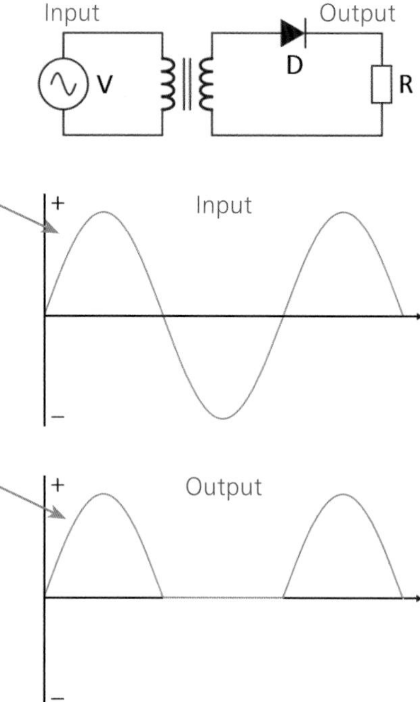

The half wave output is technically DC but very poor because only half of the waveform appears across the load. By using another diode, the other half of the waveform can be rectified.

Full wave rectification

To achieve a much better output, a transformer that is tapped at the center of the secondary windings is required. The load is now connected to this point. A second diode (D_2) is connected with its anode (+) to the transformer, and the cathode (-) to the cathode of D_1.

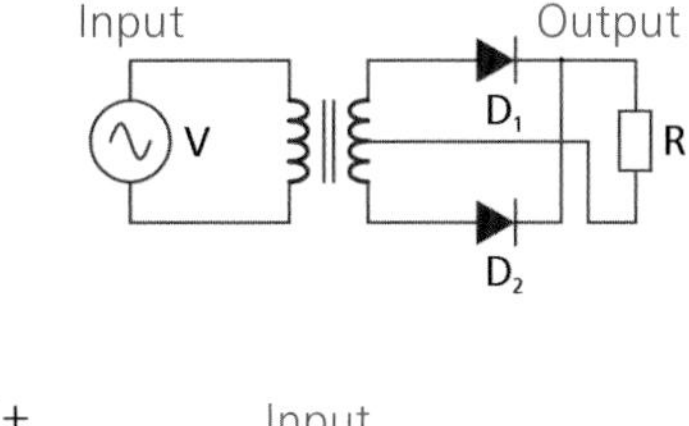

In this configuration, D_1 conducts during the positive half cycle, and D_2 conducts during the other half cycle.

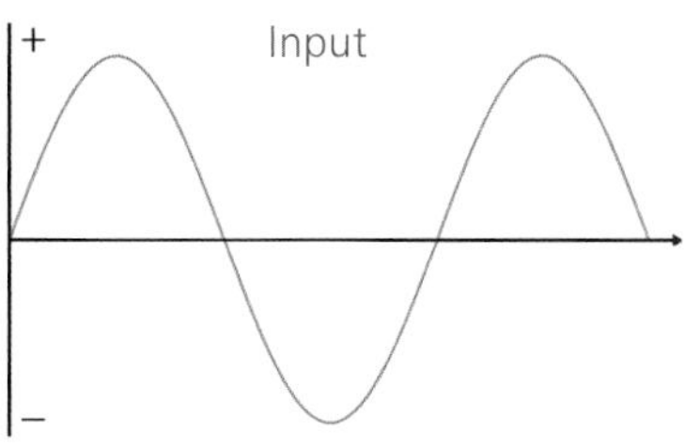

You may think that the D_2 is conducting during negative half cycles, but in fact the center tap of the transformer means that when D_1 sees a positive half cycle then D_2 sees a negative half cycle and so can't conduct.

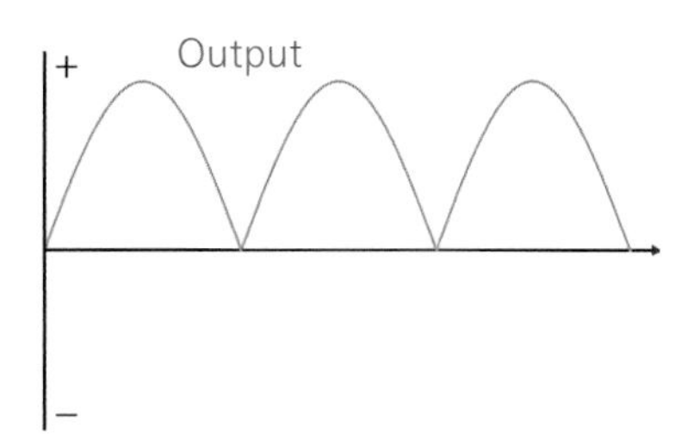

When the situation is reversed and D_1 sees a negative half cycle and doesn't conduct, D_2 now sees a positive half cycle and conducts. This may be a little confusing at this stage but all will be explained in more detail later. The important thing is that the oscilloscope will show that the gap where there was a missing half cycle in half wave rectification has now been filled in by a positive half cycle. This is full wave rectification.

Ripple

This is where DC output from rectified AC isn't fully smooth. An oscilloscope is ideal for observing this and the effects of increasing the smoothing.

See pages 89-90 for details of the full wave bridge rectification circuit.

Use a capacitor to minimize ripple in a rectifier circuit.

Safety First

Be very careful when using probes on circuits containing high voltages.

Most DC circuits you will come across are usually powered by low-voltage sources, be they power supplies or batteries, so the risk of accidentally receiving an electric shock is usually quite small. However, you do still need to take care when using test probes, etc., as some electronic components can be all too easily damaged by inadvertently shorting wires or tracks together when taking voltage readings or using oscilloscope probes.

Working with AC requires considerably more care, as the voltage source is usually the mains supply. Remember that not only is the voltage higher but the mains supply is a far greater energy source than a simple battery. DC power supplies that are mains driven are also designed to limit the amount of current they can deliver, but not so the AC mains supply.

There is a saying you should remember:

It's the volts that jolts but the mills that kills!

What this old cliché means is that although a high voltage can deliver a nasty jolt to the body if you were to accidentally touch an exposed wire or connector, it is the current behind it that will do the damage. It is generally accepted that only a few milliamps can deliver a fatal shock – it doesn't need amps.

The real hazard, then, when working with high AC voltages is clearly not from voltage alone but from the amount of current the voltage is capable of delivering. For example, 100 volts will deliver 100 milliamps through the human body – a current flow that becomes dangerous after just three seconds.

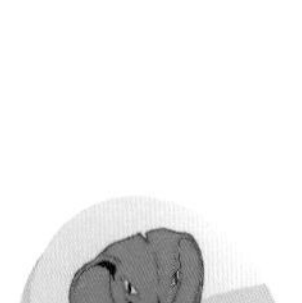

Take great care when working on circuits powered from the AC mains supply, and especially when fault-finding and repairing older equipment.

Though modern mains supplies are fitted with RCD and/or MCB devices designed to prevent you from getting a fatal electric shock if you touch something live, you should still take great care with AC:

- RCD – Residual Current Device, also called a Residual-Current Circuit Breaker (RCCB). It is designed to instantly break an electric circuit to prevent serious harm. This is the general term – it is known as a Ground Fault Circuit Interrupter (GFCI), Ground Fault Interrupter (GFI), or Appliance Leakage Current Interrupter (ALCI) in the US and Canada.
- MCB – Miniature Circuit Breaker, commonly used instead of a fuse in low-voltage electrical circuits. If tripped, it can be reset.

6 Semiconductor Principles

Learn about semiconductors – how they work and the way they are used to form a diode – and two types of transistor.

Semiconductors

It is almost time to look at diodes and transistors. To help you understand how they work you need to first know about the properties of the various materials used in their manufacture.

Electrical resistivity

Materials are grouped according to how electrically resistive they are. There are three classifications; these being conductors, semiconductors, and insulators. Because of their low electrical resistivity, metals are usually seen as good conductors.

In their normal state, semiconductors are neither good insulators nor good conductors.

The table below lists typical resistivity values for some common materials. Note that the resistivity of a material changes with temperature, so the values given are for a nominal 20°C.

Classification	Material	Typical Resistivity at 20°C
Insulators	Glass	10^{10} Ω m to 10^{14} Ω m
	Mica	≥10^{11} Ω m
	Paraffin	10^{17} Ω m
	PVC	≥10^{13} Ω m
	Rubber	approximately 10^{13} Ω m
Semiconductors	Germanium	4.60×10^{-1} Ω m
	Silicon	6.4×10^{2} Ω m
Conductors	Aluminum	2.82×10^{-8} Ω m
	Brass	8×10^{-8} Ω m
	Copper	1.72×10^{-8} Ω m
	Gold	2.44×10^{-8} Ω mV
	Mild Steel	15×10^{-8} Ω m
	Nickel	6.99×10^{-8} Ω m
	Silver	1.59×10^{-8} Ω m

The most common semiconductors are silicon and germanium.

The most common semiconductors used in electronics are silicon and germanium. As the temperature rises, their resistivity reduces until at a particular temperature they basically become conductors. Conversely, reducing the temperature of these semiconductors to well below room temperature causes their resistivity to increase to the point at which they become insulators.

Doping

A pure semiconductor contains no free electrons and so does not conduct. If the semiconductor were heated it would cause the release of a few electrons and thus allow for a very small amount of current flow.

Boron, aluminum or indium is used to make P-type material. Phosphorus or arsenic is used in the manufacture of N-type material.

Fortunately, there is a much better way of increasing current flow – by adding a small amount of impurity to the semiconductor. This process, where a few parts per million of impurities are added to a pure semiconductor, is called *doping*.

Normally, the atoms in a semiconductor are very tightly bound together by a structure of four electrons. Depending on the impurity used, two types of material can be created.

N-type: When an impurity with a five-electron structure, such as phosphorus or arsenic, is added to a semiconductor, the result is a leftover electron that has nothing to bond to.

Impurities that cause an increase in the number of free electrons are called *N-type*, and hence produce N-type material.

P-type: Conversely, adding an impurity with only a three-electron structure means that the semiconductor and not the impurity now has a free electron with nothing to bond to.

This is called a *hole*. Impurities that create holes are called *P-type*, and so produce P-type material. Boron, aluminum and indium are examples of P-type impurities.

Silicon

Pure silicon is naturally a good insulator, but doping it with a minute amount of either N-type or P-type impurity turns silicon crystal from a good insulator into a reasonable (but not great) conductor, hence the name of *semiconductor*.

In the early days of transistors, germanium was extensively used, but today silicon is the preferred semiconductor due to its abundance and affordability.

Initially, germanium was the preferred semiconductor, but today most diodes, transistors, and semiconductor chips use silicon. Silicon is now the heart of any electronic device, and is used in such quantities that the area of concentration of chip-manufacturing companies is even referred to as Silicon Valley. Anything that uses radio waves, an electronic circuit, or is computerized, depends on semiconductors.

The P-N Junction

Because of its construction, a diode allows current to flow in one direction only.

When pieces of P-type and N-type silicon are placed together you get an unexpected result that can be put to good use in a diode, as we will see later.

A P-N junction is created by joining a piece of P-type semiconductor material to a piece of N-type semiconductor material. If a hole is thought of as a positive charge, and an electron as a negative charge, then it follows that these charges will be attracted towards each other, bonding together at the junction of the two materials.

The area where the P-type and N-type materials are bonded together is called the *depletion area*.

This area where the bonding takes place is called the *depletion layer*. Eventually, when bonding is complete, the movement of holes and electrons stops and the depletion layer becomes stable. A potential difference, called the *contact potential*, now exists across the junction.

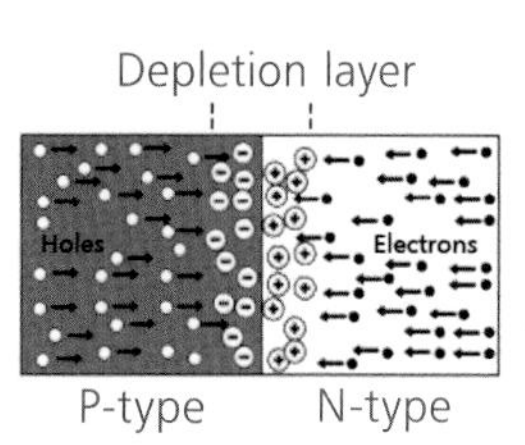

The diode

A diode is basically a P-N junction placed inside a glass or molded container, hence the terms *P-N junction diode* or *semiconductor diode*. If you apply a voltage across the junction, first in one direction then the other, and consider what happens, you can see how a diode works. This is called *biasing*.

Forward bias

This is where you apply a voltage that makes the P-type material positive with respect to the N-type material. This forces the depletion layer closer together and become thinner, thus allowing the holes and electrons to move through the junction. This results in a current flow.

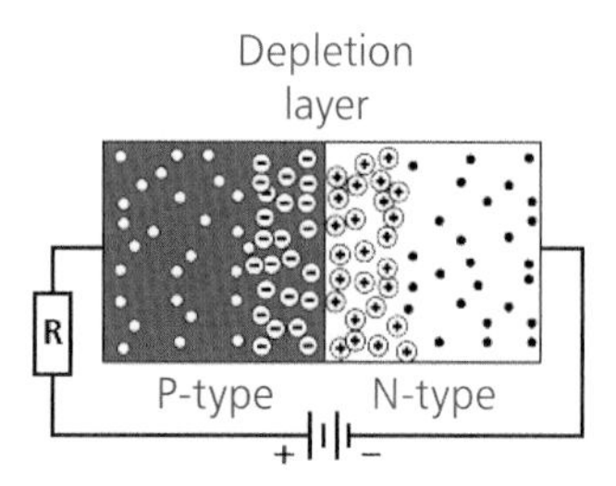

Although the rectification properties of a P-N junction were discovered in the late 1800s, the principle by which it worked was not fully understood for more than 30 years!

The amount of bias voltage depends on the type of semiconductor. Silicon requires approximately 0.7 V of forward bias to narrow the depletion layer sufficiently to allow current to flow. Approximately 0.4 V forward bias is required to cause current flow for germanium.

Increasing the forward bias further causes the material to become a good conductor, with a resulting rapid rise in current flow.

Reverse bias

If the external voltage is reversed and applied the opposite way round so that the P-type material becomes more negative with respect to the N-type material, the depletion layer becomes strengthened so that no current will flow.

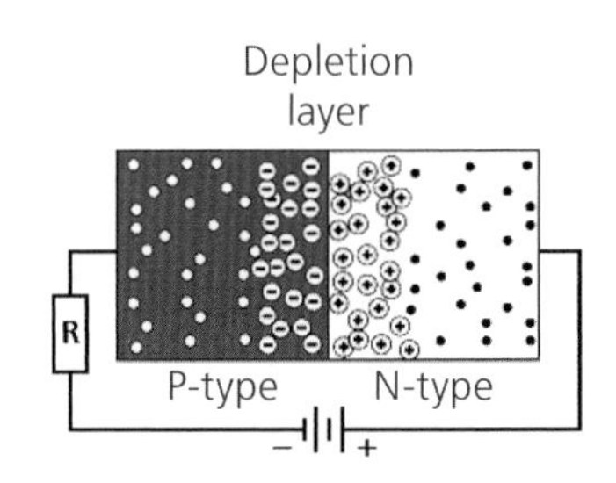

If a diode is connected the wrong way round in a circuit then no current will flow, meaning it is reverse biased.

This explanation of forward and reverse biasing makes it clear as to why the diode will only allow current to flow in one direction. This property is what allows diodes to be used for converting AC to DC as explained earlier. A diode of suitable current rating can also be used in a DC supply line to prevent equipment from being damaged if it is accidentally connected the wrong way round to the voltage source.

The cat's whisker

No, this isn't a joke – there really was a diode called a *cat's whisker*! In the early days of radio, and before semiconductors and their properties had been fully discovered, the detection of radio signals was achieved by using a type of crystal detector (the earliest type of semiconductor) that became known as the cat's whisker.

The cat's whisker detector consisted of a crystal in a fixed holder, and a thin copper wire (the cat's whisker) in a movable holder. You had to carefully position the tip of the wire on a suitable position on the crystal to create a P-N junction for it to work.

It was quite a large device compared with today's diodes and required very careful adjustment for it to work properly. Consisting of a piece of crystalline mineral – usually galena – with a fine wire touching its surface, its main function was to act as a rectifier for the early radio receiver. When properly adjusted, the diode rectified the radio frequency being received and provided sufficient DC voltage to power headphones. Listening to the radio in those early days was nowhere near as simple as it is today!

The crystal set

Very simple radio receivers consisted of just a few inexpensive components – often homemade – and required no external power source, as the rectifying action of the crystal detector provided the power, small as it was. They became known as *crystal sets*.

Being cheap and reliable, crystal radios were the first widely used type of radio receiver and were a major driving force in the introduction of radio to the public, especially when radio broadcasting began around 1920.

The rectifying property of a contact between a mineral and a metal had been discovered in 1874. The crystal diode was the forerunner of today's diode, and it wasn't until 1920 that the crystal set started to make way for the first amplifying receivers.

The PNP Transistor

The transistor makes use of everything you have learned about the P-N junction, and is simply three pieces of semiconductor material connected in a way that lets you control the flow of current. How the P-type and N-type material is connected together is obvious from the name.

The transistor is considered to be one of the greatest inventions of the 20th century, and today over a billion individual transistors are manufactured every year.

A brief history

The first practical point-contact transistor was invented in 1947 by American physicists John Bardeen, Walter Brattain, and William Shockley at the Bell Laboratories in the USA. Shockley announced a much improved bipolar junction transistor some three years later, and soon the transistor quickly started replacing valves in electronic circuits. The transistor was such an important invention and milestone in electronics that in 1956 the three were jointly awarded the Nobel Prize for physics.

PNP configuration

The PNP transistor consists of a thin layer of N-type material sandwiched between two pieces of P-type material. It has three connections: those to the two P-type materials are called the *emitter* (E) and *collector* (C), whilst that to the N-type material is called the *base* (B).

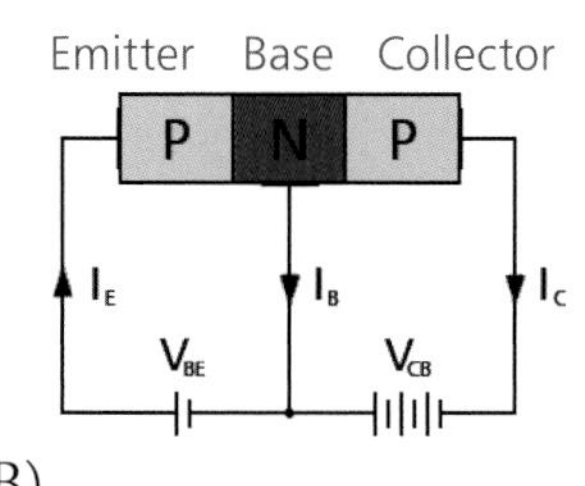

When the transistor is connected to an electrical source as shown, the base-emitter junction becomes forward biased, whilst the base-collector junction is reverse biased. Connected in this way, it is possible to control the output from the collector. Note that the base-emitter voltage source is smaller than the collector-base voltage source.

A bipolar PNP Transistor will ONLY conduct if both the collector and base are negative with respect to the emitter.

Circuit symbol

Shown opposite is the circuit symbol for the PNP transistor. As this is a standard symbol it is not always necessary to include the C, B, and E letters, as the connections are easily identified.

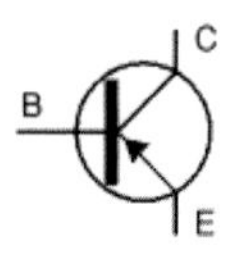

Basically, the construction of this transistor is like having two diodes connected back-to-back so that they share a common N-type material. The arrow that identifies the emitter terminal points inward to show that a PNP transistor "sinks" current into its base. In this device, holes are the more important carriers.

The NPN Transistor

Like the PNP transistor, the NPN device is also a bipolar junction transistor, or BJT. Initially, the PNP type was the most widely used, but today the NPN configuration is the most common.

Always ensure that the correct polarity is observed when connecting a transistor to its power source to avoid potential component failure.

NPN configuration

The construction is very similar to the PNP transistor except that, as the name suggests, the NPN transistor consists of a thin layer of P-type material sandwiched between two pieces of N-type material.

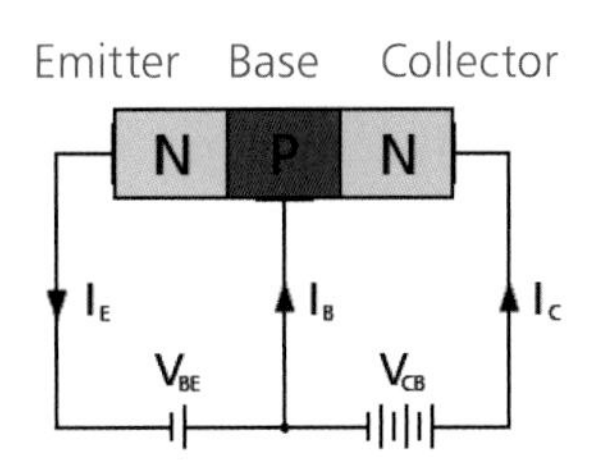

It, too, has three connections – collector (C) and emitter (E) for the N-type materials, and base (B) for the P-type.

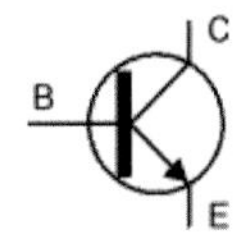

When the NPN transistor is connected to a power source in the opposite way to a PNP transistor, the N-P (base-emitter) junction is forward biased and the P-N (base-collector) junction reverse biased, making it once again possible to control the output from the collector.

Individual transistors are quickly being superseded by the integrated circuit containing hundreds of transistors and other components on one single chip.

Current devices

As with the PNP type, the NPN transistor is also a three-layer device constructed from two semiconductor diode junctions joined back-to-back: one forward biased and one reverse biased. The NPN transistor requires the base to be more positive than the emitter, whilst the emitter has to be more positive than the base for the PNP type.

Bipolar transistors are therefore current-operated devices. The amount of collector output current can be controlled by the amount of current flowing into the base connection.

Switching

In an NPN transistor, the flow of current is from the collector to the emitter. A PNP transistor switches ON when there is no flow of current at the base terminal of the transistor.

In a PNP transistor, the flow of current is opposite and runs from the emitter to the collector. Consequently, a PNP transistor is switched ON by a low signal, whereas an NPN transistor is switched ON by a high signal. It "sources" current through its base.

A transistor can be used as a switching device or an amplifier.

Semiconductor Roundup

Here is a summary of what you have discovered about semiconductors and their properties. Please take note of these points as this will help you understand the function and purpose of the wide variety of transistors (and diodes) introduced in the following two chapters.

The transistor does have its limitations in high-power situations, and is not as robust as the valve (or vacuum tube) that it has replaced.

- There are just two kinds of bipolar sandwich – PNP and NPN. This is the basis for transistors.
- Transistors are semiconductor devices made using germanium or silicon. The PNP transistor is biased in the opposite way to the NPN transistor, so consequently the flow of current in the NPN device is in the opposite direction to the current flow in the PNP device.
- In a PNP transistor, the flow of current runs from the emitter to the collector as indicated by the inward-pointing arrow in its circuit symbol.
- In an NPN transistor, the flow of current runs from the collector to the emitter as indicated by the outward-pointing arrow in its circuit symbol.

The transistor can be thought of as two diodes connected back-to-back.

- A PNP transistor is made up of two P-type material layers sandwiching a layer of N-type material. The NPN transistor is made up of two N-type material layers sandwiching a layer of P-type material.
- In an NPN transistor, the **collector** has to be connected to a positive voltage to allow a flow of current from the collector to the emitter.
- For a PNP transistor, a positive voltage has to be connected to the **emitter** to generate a flow of current from the emitter to the collector.
- The working principle of an NPN transistor is that an **increase** in the base current will switch the transistor ON. If you **reduce** the base current, the transistor switches OFF.
- The working principle of a PNP transistor is that when you **decrease** the base current it will cause the transistor to switch ON. When you **increase** the base current, the transistor switches OFF.

Being very small, a large number of transistors can now be manufactured as a single integrated circuit.

7 Diodes

There are many types of diodes, all designed for a specific function or application. This chapter explains the more common ones you will encounter in electronics, how they work, and where they are used.

Diode Basics

A diode conducts current in one direction only if it is forward biased.

The diode is a common device used in a wide variety of electronic circuits. From Chapter 6 you will remember that it is an electronic device that allows current to flow in one direction only. The current flows in the forward direction, but if the diode is connected in reverse then it acts as a blocking device and no current will flow.

All diodes are made using a P-N junction, with the two connections known as the *anode* and the *cathode*. The P material is the anode and the N material is the cathode. When the anode is sufficiently positive with respect to the cathode, current will flow from anode to cathode as indicated by the arrow in the diode symbol.

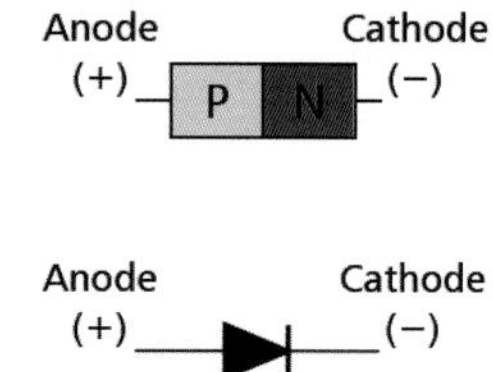

Special purpose diodes

By now you should have understood how a diode works. The current-voltage characteristics of a semiconductor diode can be tailored through selection of the semiconductor material and precise quantities of the doping impurities introduced at the manufacturing stage. In this way, special purpose diodes can be created.

By convention only the cathode is marked on the body of a diode. When connected in a circuit this should point to the negative side of the supply.

Below is a shortlist of some of the more common special function diodes you will come across and their purpose.

Diode	Purpose
LED	To emit light by release particles of light energy called *photons*
Zener	To regulate and control voltage
Varactor/varicap	To electronically tune television and radio receivers without using moving parts
Gunn/tunnel	To generate accurate and stable radio frequency oscillations
Avalanche	To protect electronic circuits from high-voltage surges or spikes

Rectifier

Although all diodes provide a rectifying function, the term *rectifier* is normally reserved for diodes intended solely for power supply application, such as a *bridge rectifier*. This differentiates them from low-power diodes intended for small signal circuits.

Light-Emitting Diode

There are a number of different types of diode designed to do specific jobs. The one you will probably recognize the most is the LED or light-emitting diode. The LED has many uses, such as the On or Standby indicator light on televisions and DVD players. The symbol for the light-emitting diode is shown here.

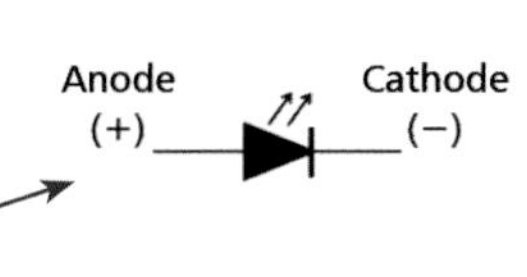

It is also used in most remote control handsets, though this time it uses invisible light to send signals to the appliance.

Unlike the miniature light bulb it replaces, the LED has no filament to burn out, uses only a very small amount of electrical power to generate light, and is much cheaper to produce. It is also available in different colors and sizes.

You will notice in the image below that because the LED has a plastic body it can be molded into different shapes, which also makes it versatile. Remember, though, that the color emitted by an LED is determined by the semiconductor material and not the plastic body, which is only colored to help identify what color the diode is designed to emit. Some can emit two or more colors.

Light-emitting diodes are available in red, orange, yellow, green, amber, blue, and white. Blue and white LEDs are more expensive than the other colors.

Photons

Just like a normal diode, the light-emitting diode consists of a P-N junction formed from a particular semiconductor material that releases particles of light energy called *photons* when the electrons and holes recombine across the depletion layer.

LED technology is advancing at such a pace that the light-emitting diode is rapidly taking over from the traditional filament lamp. Very low wattage LED lamps are replacing light bulbs around the household, cutting electricity bills whilst producing high light output. Motor cars now leave the production line equipped with LED lights as standard, and traffic lights, too, are now using LEDs.

The color of the light emitted by an LED corresponds to the energy of the photons, and is determined by the energy (or voltage) required for electrons to cross the depletion area.

Recent technology has resulted in very high-energy white light being produced using a combination of yellow and blue light. This appears white to the eye. This type of LED is rapidly taking over from the filament lamp.

Zener Diode

The main function of a Zener diode is as a voltage reference device.

A *Zener diode* is a diode that makes specific use of the reverse bias effect, but in a special way. It is manufactured to actually start conducting when the reverse bias voltage is increased to a specific point. The correct term for the Zener is *reverse breakdown diode* – it is the effect that is called *Zener breakdown*.

The voltage at which a diode begins to conduct when it is reverse biased depends on the amount of semiconductor doping used. As the doping is increased, the breakdown voltage drops. It is possible to accurately specify the breakdown point by precisely controlling the doping for a particular diode.

Breakdown voltage

Zener diodes are manufactured with breakdown voltages ranging from 2.7 V to over 150 V, though, strictly speaking, a Zener diode has a breakdown voltage of below 5 V, and the *avalanche diode* is used for breakdown voltages above that value.

This is the circuit symbol for a Zener diode.

Anode (+) Cathode (–)

Simple voltage regulator

With the addition of a suitable resistor, a Zener diode can be used to provide stable voltage regulation. The very simple circuit is shown opposite.

+ R – Supply voltage Zener diode Regulated voltage

A series resistor should always be used to limit the breakdown current and thus avoid destroying the Zener diode by overheating.

The Zener diode is always operated in its reverse biased condition, and helps to maintain a specific DC voltage output as the current being drawn changes, or in spite of any variations in the supply voltage.

The Zener voltage regulator consists of a current-limiting resistor (R) connected in series with the Zener diode in the reverse biased condition across the supply voltage. The regulated output voltage is selected to be the same as the breakdown voltage of the diode.

Zener diodes are listed primarily by current rating, and secondly by breakdown voltage. For general electronic circuits, a typical Zener diode is the 1.3 W, BZX85 series or the smaller 500 mW, BZX55 series. The breakdown Zener voltage is usually given as, for example, for a 5.6 V diode, C5V6. The full diode reference for the smaller Zener would therefore be BZX55C5V6.

Varactor or Varicap Diode

Varactor

This is basically a voltage-controlled capacitor, and for simplicity it can be thought of as a diode and capacitor in one small package. This is evident from its circuit symbol shown opposite. Despite having no moving parts, the capacitance is actually variable over a small range using a controlled voltage source.

The circuit symbol for the varactor or varicap diode combines the diode and capacitor symbols to make it clear that its function is that of a variable capacitor rather than a rectifier.

Operation

When learning about AC circuits you saw that there is always some stray value that stops something from being "pure". This is true of the P-N junction. All diodes exhibit some capacitance at the depletion area, and as the size of the depletion layer changes depending upon how it is biased, so does the capacitance.

Varactors are manufactured to exploit the above effect and to increase the capacitance variation. The varactor is always operated in a reverse-biased state so that no DC current can flow through it. Because the amount of reverse bias controls the thickness of the depletion layer it will, therefore, also control the varactor's junction capacitance.

The varactor can thus be used as a small solid state variable capacitor. The capacitance is inversely proportional to the square root of the applied voltage, so it is possible to calculate the capacitance for a given reverse bias. Typical working reverse bias is from 2 V up to 20 V. Up to 60 V is not unknown, although at the very top end of the range little change in capacitance is seen.

When operated in a circuit it is necessary to ensure the varactor diode remains reverse biased to function correctly.

Varicap diode

This is simply another name for the varactor that was developed in mid-1960. The name was trademarked as "Varicap" in 1967.

Uses

The varactor/varicap is used in a radio or TV tuned circuit, usually in parallel with any existing capacitance or inductance. A small change in the reverse bias voltage causes a capacitance change that accurately alters the tuned frequency. Three leaded devices are generally two common cathode connected varicaps in one package.

Avalanche Diode

In electronics, the avalanche diode is a diode that is designed to break down and conduct at a specified reverse bias voltage, similar to the breakdown seen in a Zener diode, but with a difference.

Avalanche diodes are designed to break down at a well-defined reverse voltage without being destroyed by heat.

Construction

In a typical diode or other semiconductor, when avalanche occurs it usually causes a catastrophic failure of the component due to excessive heating.

The junction of an avalanche diode is designed such as to prevent the concentration of current flow causing hot spots. This means that the diode remains undamaged by the breakdown. You will notice that its similarity to the Zener is mirrored in the circuit symbol shown below.

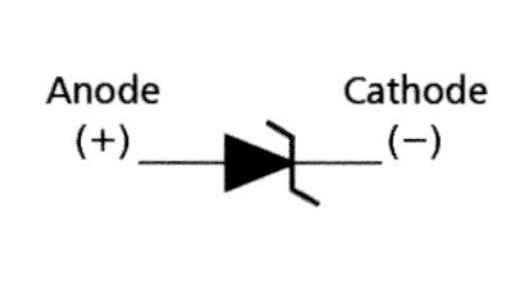

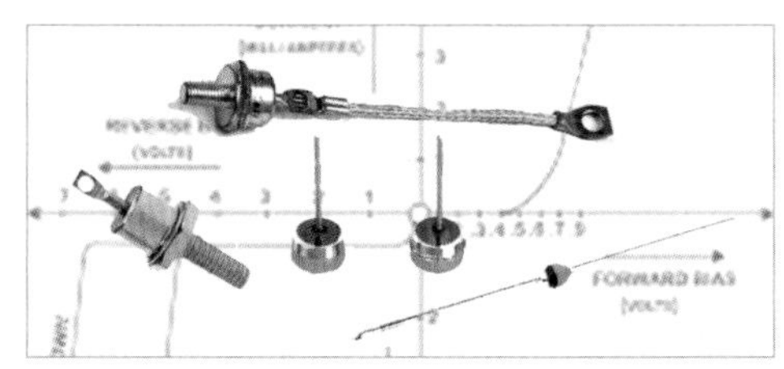

Unlike Zener diodes, which are quite heavily doped and thus the width of the depletion region is very thin, avalanche diodes are only lightly doped. Because of this thin depletion region or layer, reverse breakdown occurs at lower voltages in the Zener diode.

However, the depletion layer is wider for the lightly doped avalanche diode and hence reverse breakdown occurs at higher voltages. The actual breakdown voltage of a particular avalanche diode can be carefully set by controlling the doping level during the manufacturing process.

Hot tip

Avalanche diodes are often used for protecting circuits against transient high voltages that otherwise would damage the circuit.

Uses

Avalanche diodes can be used in the following applications:

- As white noise, radio, and microwave frequency generators.
- For protecting circuits against damaging or high voltages.

Disadvantages

- Produce more radio noise than the normal diode.
- Much higher operating voltage may be needed.
- The output is not linear because of the avalanche process.

Photodiode

A *photodiode* is actually the opposite of a light emitting diode. The LED converts an electric current into light by releasing photons, whereas the photodiode is a semiconductor device that converts light into an electrical current by absorbing photons.

To aid their sensitivity, photodiodes may contain optical filters or built-in lenses, and may have large or small surface areas. For example, the traditional solar cell used to generate electricity from solar power is a *large area photodiode*. It also has other names – *photo-sensor*, *photo-detector*, or *lightdetector*.

A photodiode is a semiconductor device that converts light into an electrical current.

Construction

Like most of the special diodes covered so far, the photodiode is similar to a P-N junction semiconductor device that is designed to operate in the reverse bias condition in a specific way. However, in contrast to an ordinary P-N diode, it contains an additional layer between the P and N materials, hence the term PIN diode.

PIN diode

This type of diode has a wide, undoped intrinsic semiconductor region between the P-type and N-type semiconductors. The P-type and N-type materials are typically heavily doped to keep resistance low so as not to impede current flow. The wide intrinsic region is pure or undoped semiconductor, also called *i-type semiconductor*. The PIN structure provides very fast response times, so PIN diodes are mostly used in high-speed applications.

Operation

The way a photodiode works can be simply explained. A normal P-N junction diode allows a small amount of electric current flow under reverse bias conditions. To increase the current flow it is necessary to generate more carriers. The photodiode introduces photons (light electrons) to generate extra charge carriers in the depletion region. The more light, the greater the current flow.

The circuit symbol identifies the photodiode by the two arrows pointing into the diode, and not away from it, as for an LED.

The very recently developed solaristor (from SOLAR cell transISTOR) is a compact two-terminal self-powered phototransistor.

...cont'd

All semiconductor devices such as diodes, transistors, and ICs contain P-N junctions, and will not function correctly if they are accidentally exposed to unwanted electromagnetic radiation.

Types of photodiodes

Different types of photodiodes have been developed based on specific application, though the working operation of all the types is the same. For example, PIN photodiodes were developed for high-speed application by increasing the diode's speed of response.

The different types and their advantages or disadvantages are listed in the following table:

Type	Advantages/Disadvantages
P-N Junction Photodiode	• simple device • low cost • small size • long lifetime • generates a low level of noise • low sensitivity to temperature changes
PIN Junction Photodiode	• high response speed • wide bandwidth • high quantum efficiency – (QE is the ratio of the number of carriers collected by the solar cell to the number of photons of a given energy falling on the solar cell)
Avalanche Photodiode	• high sensitivity to light • larger gain Disadvantages • generates higher level of noise than the P-N photodiode • high reverse bias voltage required to achieve avalanche

A *phototransistor* is a light-sensitive transistor. Also called a *photobipolar transistor*, it is simply a bipolar transistor encased in a transparent case so that light can reach the base-collector junction.

Of the three photodiodes, the P-N junction photodiode and PIN junction photodiode are the most widely used.

P-N junction photodiodes were the first form of photodiodes and were widely used before the development of PIN photodiodes. Today, they are not widely used.

Some early OC series germanium transistors were encased in clear glass and covered with a paint-like coating. It was possible to scrape away the coating to expose the junction and use these devices as simple photodiodes instead of buying specific ones.

Schottky Diode

The *Schottky diode*, also known as *Schottky barrier diode* or *hot-carrier diode*, is a semiconductor diode formed by the junction of a semiconductor with a metal. It has a low forward voltage drop and a very fast switching action and is named after the German physicist Walter H. Schottky.

Construction

Instead of a semiconductor to semiconductor junction as in a conventional diode, the Schottky diode uses a metal to semiconductor junction to create what is called a *Schottky barrier*.

The N-type silicon semiconductor is retained, but typical metals such as aluminum, platinum, molybdenum, chromium, or tungsten replace the P-type semiconductor. The metal acts as the anode, and the N-type material acts as the cathode, meaning that conventional current can flow from the metal side to the semiconductor side but not in the opposite direction. Properties of this Schottky barrier are very fast switching and low forward voltage drop.

Operation

When sufficient forward voltage is applied, a current flows in the forward direction. A standard silicon P-N junction diode has a typical forward voltage of 600–700 mV, while the Schottky diode has a much lower forward voltage of 150–400 mV. This lower forward voltage results in higher switching speeds and allows for better circuit efficiency. The Schottky diode also produces less unwanted electrical noise than P-N junction diode.

The Schottky diode has a low forward voltage drop and a very fast switching action.

Uses

Schottky diodes have a variety of applications:

Anode (+) Cathode (−)

- As general purpose rectifiers.
- Widely used in power supplies.
- For signal detection.
- In radio frequency (RF) circuits.
- Widely used in logic circuits.
- Where immediate ON to OFF switching is required.

Tunnel/Gunn Diodes

Although these two diodes are not common in everyday electronics, they are included here for completeness. They are both specialist devices – particularly the Gunn diode – and are a variation on the standard P-N semiconductor diode, and developed to perform a specific role in a circuit.

A tunnel diode is also called an *Esaki diode*.

Tunnel diode

Anode (+) Cathode (–)

A *tunnel diode* is a heavily doped P-N junction diode in which the electric current decreases as the voltage increases; the electric current being caused by an effect called *tunneling*, hence the name.

Construction

The P-type and N-type semiconductors are heavily doped, meaning that a large number of impurities are introduced into the materials. The concentration is 1000 times greater than the normal diode and so produces an extremely narrow depletion region.

This construction causes the diode to exhibit negative resistance, meaning that the current across the tunnel diode decreases when the voltage increases, unlike an ordinary diode that produces electric current only if the applied voltage is greater than the "built-in" voltage of the depletion region.

This property was discovered by Leo Esaki in 1973. He received the Nobel Prize in physics for discovering this tunneling effect, and the tunnel diode is also known as the *Esaki diode*.

Because of their low voltage operation, Gunn diodes are ideal for use as microwave frequency generators for very low-powered (a few milliwatt) microwave transceivers called *Gunnplexers*. They were first put to this use by British radio amateurs in the late 1970s.

The tunnel diode is used in high-frequency oscillators (such as in FM receivers), as logic memory storage devices, as an ultra high-speed switch, or a very fast switching device in computers.

Gunn diode

Also known as a transferred electron device (TED), the *Gunn diode* is a form of diode used in very high-frequency electronics. This two-terminal passive semiconductor component employs negative resistance and operates using the "Gunn effect" discovered in 1962 by physicist J. B. Gunn.

Because of their low operating voltage, Gunn diodes are used in radar detectors, microwave technology, and millimeter-wave and sub-millimeter-wave radio astronomy receivers.

Rectifier Diodes

Although it is just a diode, the term *rectifier* is used to describe those devices designed purely for rectifying AC to DC. As you have seen, all P-N junction diodes will do this, but often they will be required to pass high levels of current, such as when used in power supplies. Consequently, rectifier diodes are available in a wide range of current and/or voltage ratings.

The thyristor is commonly used in place of diodes to create a circuit that can regulate the output voltage.

Rectification

The process of rectification means taking an alternating current and converting or straightening it into direct current. Rectification is actually the most popular application of the diode, so it is not surprising that there are so many different types of rectifier diodes in manufacture. Rectification is necessary because the readily available mains supply is high-voltage AC, whilst electronic circuits, devices and equipment are usually designed to operate from a low-voltage DC supply.

Rectification methods

You briefly learned in Chapter 5 that there are a number of ways in which AC can be converted to DC using one or more diodes. In each case, though, and depending on the nature of the alternating current supply and the arrangement of the rectifier circuit used, the output voltage may well require additional smoothing to produce a uniform steady output that is true DC. This issue is covered later on pages 102-104.

There are three basic rectification methods:

- Half wave (using 1 diode only).
- Full wave (using 2 diodes).
- Bridge (using 4 diodes).

Before the silicon semiconductor rectifier was developed, valve (or vacuum tube) thermionic diodes and copper oxide- or selenium-based metal rectifier stacks were in common use.

Circuits for half and full wave rectification are shown on pages 68 and 69, together with the corresponding output waveforms. It can be seen that the full wave method is more efficient than half wave rectification, but in both circuits the diodes have to be of sufficient rating to safely pass the required current.

Commonly used diodes are the 1N400x series, shown above. Typically, the 1N4001 is rated at 50V 1A but for many applications, rectifier diodes with much higher current ratings are required – for example, such as in a 10 A bench power supply.

Bridge rectifier

A diode bridge is an arrangement of four diodes in what is called a *bridge circuit configuration*. When used for conversion of an AC input into a DC output, it is known as a *bridge rectifier*.

A bridge rectifier can also be configured in a circuit to act as a voltage-doubling rectifier.

Construction

Because the bridge rectifier circuit is so frequently used, specialist bridge rectifier devices are available that avoid the necessity of having to wire four diodes into the bridge configuration. Three typical types are shown here – their different sizes give an indication of their higher power-handling capacity.

A bridge rectifier circuit does exactly the same function as the full wave rectifier circuit but with advantages. Because four diodes are used, the current-carrying load on each diode is shared. Also, the transformer does not require a center tap, resulting in lower weight and cost.

Operation

The bridge rectifier circuit is shown here. Being simpler and more efficient than the full wave rectifier circuit despite requiring two additional diodes, the bridge rectifier has become extremely popular. The output waveform is exactly the same as for the full wave rectifier.

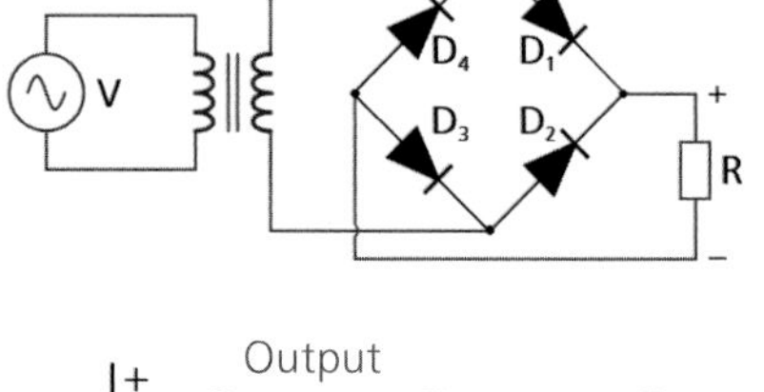

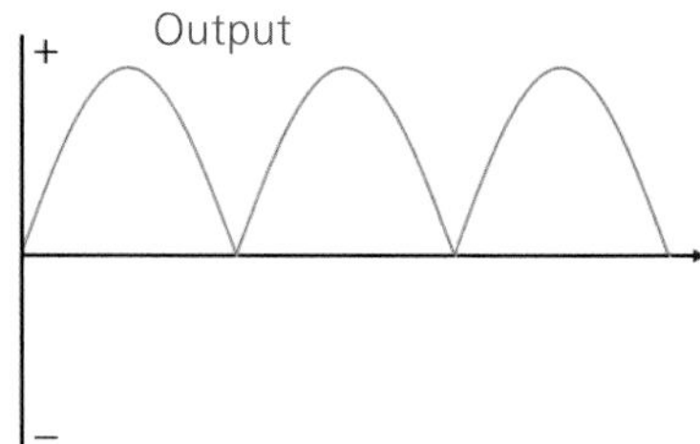

Single-phase rectifiers are commonly used for power supplies for domestic equipment. However, for the majority of industrial and high-power applications, three-phase rectifier circuits are normal.

The diodes D_1 to D_4 are arranged in series pairs, so only two diodes conduct current during each half cycle.
During the positive half cycle of the supply, diodes D_1 and D_3 conduct, while diodes D_2 and D_4 are reverse biased. During the negative half cycle of the supply, diodes D_2 and D_4 conduct whilst D_1 and D_3 are now reverse biased. Therefore, positive current flows through the load R during both half cycles of the AC input.

8 Transistors

See how to identify different transistor types and their pin layout, use of heatsinks, and testing a transistor.

Transistor Basics

Transistors are called *active components* because they introduce power into a circuit.

All individual basic components used in electronics are called *discrete devices*. These electronic components are then categorized into two general types of main components: passive and active.

Passive

These are components such as resistors and capacitors that consume or store energy, but cannot introduce any form of energy back into a circuit.

Active

Components like transistors and integrated circuits are called *active components* because they can inject power or introduce something such as amplification into a circuit.

Gain

Gain is a measure of the increase in signal at the output compared with the original input signal level.

In Chapter 6 you learned how PNP and NPN transistors are constructed and biased for operation. In practice, a transistor is extremely useful because of the ability to use a small signal applied between one pair of its terminals to control a much larger signal at another pair of its terminals.

When appropriately biased, a transistor can produce a stronger output signal of current or voltage that is proportional to a much weaker input signal. In other words, it can act as an amplifier; a property that is called *gain*.

Switching

A transistor can also be used as an electrically controlled switch to turn current on or off in a circuit. In this case, the transistor is biased to either fully conduct or turn fully off. The current it switches on or off does not pass through the transistor but through another device, such as the contacts of a relay.

A transistor can be used in an analog or digital mode. When biased to only turn on or off, it is called a *switching device*.

The transistor simply needs to pass enough current to energize the relay; the benefit being that the switched current and the switching transistor are isolated from each other. Also, the switched current can be large, whilst the switching current can be very small. A simple switching circuit is shown where a NPN transistor is used to operate a relay that can switch a higher current.

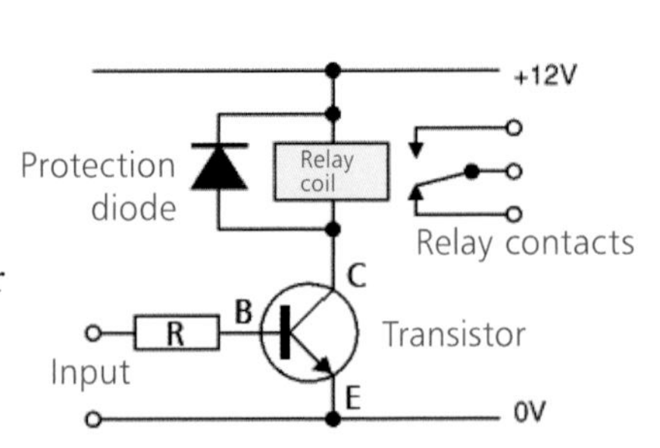

Base resistor

If you look at the switching circuit you will notice the series resistor R between the input and the base of the transistor. This is called a *base resistor*, and its function is to limit the current flowing into the base.

You will remember that a transistor is really just like two interconnected diodes, and to turn the base-emitter diode ON then a forward bias of only about 0.6 V is needed. Any greater voltage would mean more current. Many of the smaller transistors are rated for a current flow of no more than 100 mA; sometimes less. A base resistor of suitable value should always be used to protect the transistor from blowing up if allowed to draw more current than its maximum rating.

The value of the resistor and voltage across it will set the current flowing through it, as per Ohm's law. The resistor needs to be small enough to allow sufficient current to feed through to the base, whilst being large enough to effectively limit any excessive current. Check the transistor's datasheet to see how much current it requires – typically, up to 10mA will usually be sufficient.

Note that a current-limiting resistor should also always be used with an LED to avoid the diode suffering the same fate.

How a transistor operates in a circuit is controlled by how its base is biased.

Always use a base-biasing resistor to avoid a transistor failing due to being made to pass too much current.

Other applications

Transistors are used in oscillator circuits for producing a periodic signal that swings between a high and low voltage. They can also be combined to create the fundamental AND, OR, and NOT logic gates.

Mechanical construction

Like diodes, transistors come in a variety of shapes and sizes depending on the work they have been designed to do and their current-handling capacity. With so many varieties and different pin configurations, identifying the leads on a particular device can be quite confusing. Over time, many differently shaped cases have also been developed so you will need to know these details too.

Always make sure you correctly identify the pin arrangement for a particular transistor.

Case Identification

There are thousands of transistors now being produced in a wide variety of packages. Make sure you have the correct one, especially with the TO92 series.

Although transistors only have three leads it is very important that they should be correctly connected in a circuit. If not, the transistor could be damaged the moment the power to the circuit is switched on.

With just three leads you would be forgiven for thinking that this shouldn't be difficult, but this is far from true. Because there are now so many different types of transistors cased (or packaged) in a wide variety of materials of various shapes and sizes, it is very easy to make a mistake when working out which lead is the collector, the base, or the emitter.

Case (or packaging)

Just a few of the more common transistor cases are shown here, but as you can already see, the popular TO92 style is available with three different C B E configurations. This is true of many other types.

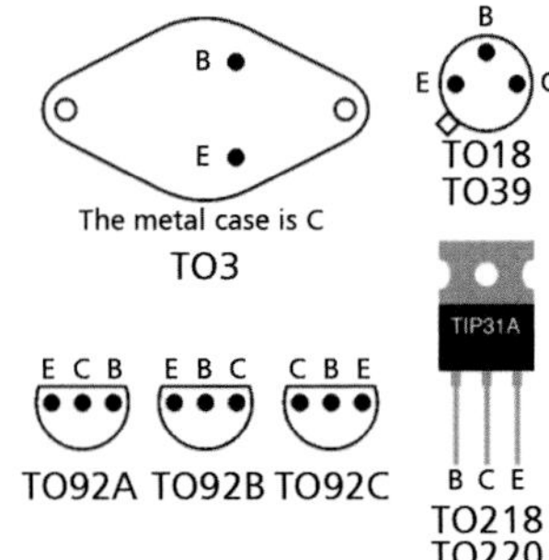

Viewed from below, leads pointing towards you

Note that transistor lead configurations are identified by viewing the base of the device, unlike logic chips that are identified when looking down on the device from above.

Common transistor packaging

The case of a transistor is often referred to as its *packaging* – for example, a BC109 is available in a TO18 package, whilst a 2N3055 is in a TO3 package. Transistor coding and how to initially identify their particular use will be covered shortly.

Careful identification of a transistor cannot be over-stressed, because some of the common packages are also used by other types of devices.

Here are some case types that you are most likely to encounter:

- TO92 series – widely used style of package. Case is made of plastic or epoxy. Offers compact size at a very low cost. Transistor code is marked on the body.

- TO18 and TO39 – metal can-type of package, fast becoming obsolete. Transistor is hermetically sealed to protect from moisture and contamination. TO39 is the larger version of the TO18 package.

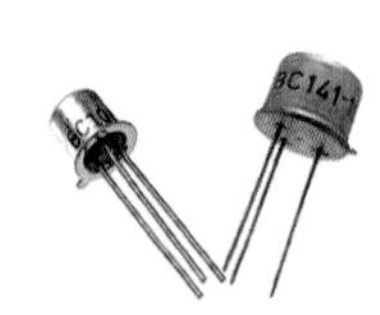

- TO218 and TO220 – a small plastic molded package with a flat metal tab at the back. Used for transistors or devices with two to seven leads. The primary difference between TO218 and TO220 is in the shape of the tab, as shown. It serves as a heatsink (see page 96) but also allows the transistor to be mounted onto a larger heatsink if required. As the tab may be internally electrically connected to the device, an insulating pad is often needed to avoid any shorting.

A simple piece of thick aluminum screwed to the tab makes a suitable cheap heatsink for the TO218 and TO220 series.

- TO126 – plastic molded outline package smaller than TO218 and TO220. Normally has a metal plate at the back for mounting to a heatsink if required. Good power-handling capability; can handle more power than the TO92.

- TO226 – small type of plastic molded three-terminal package commonly used for discrete devices such as transistors, thyristors, and voltage regulators. The back is semi-circular, while the device code or number is marked on the flat front face. The three leads protrude from the bottom of the package. Structure is very similar to the TO92 but with a longer body.

- TO3 – all-metal package that is capable of high-power dissipation. The semiconductor chip is mounted directly on the base of the can then covered with a metal cap. A typical TO3 package has two protruding terminals: one for the base and the other for the emitter – the body itself is the collector. Advantages include excellent power-handling capability, ease of mounting, and high durability.

- TO66 – smaller version of the TO3 metal package, offering high-power dissipation. Holes at either end of the package allow for mounting on a heatsink to increase power capability.

Always use insulating mountings with TO3 and TO66 transistors, as the metal body is the collector. Various mounting kits are readily available for this.

Heatsinks

A transistor does not necessarily require a heatsink if it is being operated well within its maximum current limit.

A *heatsink* is used to increase the power-handling capability of a transistor. Made of a heat-conducting metal, fins on the heatsink aid in dispelling this heat. The TO3 package is designed for fitting to large heatsinks to increase its power-handling capability. The transistor and heatsink may be electrically isolated from each other using mica or ceramic pads.

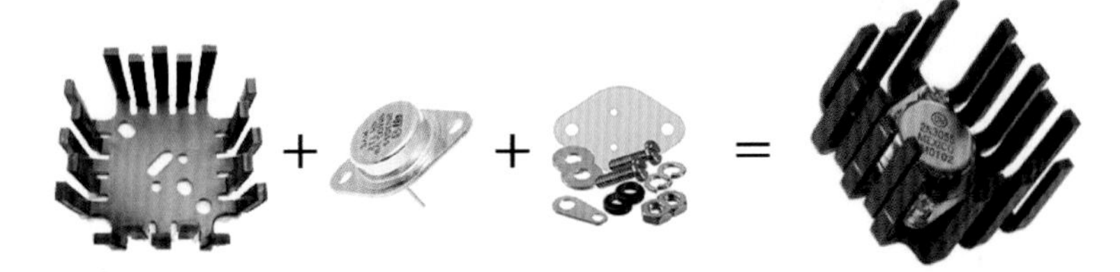

Heatsink construction

There are many designs for heatsinks, but the majority typically comprise a base section and a number of protrusions in the shape of fins or rods/fingers. The device to be cooled is attached to the base. Heat is conducted through the base into the protrusions that then transfer the heat to the air.

Typically, heatsinks are constructed from aluminum or copper. Despite being more expensive and heavier, copper is preferred because it has a very high thermal conductivity. This means that the rate of heat transfer through a copper heatsink is also very high. Although the thermal conductivity of aluminum is lower than that of copper, it is still high, plus it has the added benefits of being lighter and cheaper, making it useful where weight is a major concern.

Always use a smear of thermal conductive paste to ensure good heat transfer when mounting transistors on a heatsink.

Thermal interface

For maximum heat transfer from device to heatsink it is important that the two make very good thermal contact. Although manufactured to have very flat and smooth surfaces, electronic components and heatsink bases in fact have rough surfaces if inspected under a microscope. This means there are very few points of good contact, but lots of tiny air gaps.

Because air has a low thermal conductivity, conduction of heat from the device to the heatsink will be poor, so it is important to use a thermal interface material (TIM). This is like a paste that is applied to the base of the heatsink to fill these gaps and provide much better heat conduction between the device and its heatsink.

Lead Identification

Earlier, you saw that although some transistors share the same package shape, their pin-out arrangement may be completely different. The table below is more comprehensive than what has been covered so far, but is by no means exhaustive – there are considerably more, but these are some of the common ones you are likely to encounter in many electronic circuits.

Take care when fault-finding on a circuit, as some transistor packages are also used by other types of devices. Always check the device code to confirm that a suspected device is, in fact, a transistor before replacing it!

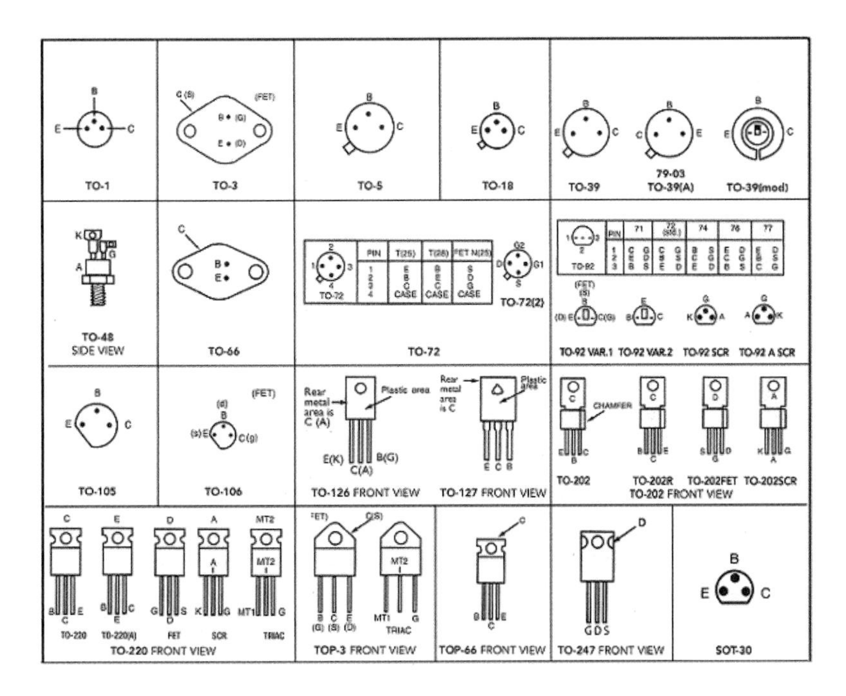

Datasheets

There are literally thousands of different transistors. Many of them have similar characteristics, so it is often possible to replace a faulty transistor with a slightly different one, if only to test if the circuit will start working again.

Manufacturers provide datasheets for each device they produce, be it a transistor, diode, or other device. Therefore, it is easy to check on the characteristics, pin configuration, and usage for a particular device. The internet is invaluable for looking up datasheets – just "Google" the device code and see how many datasheet sources there are. Most are available for download in PDF format.

Remember you can only replace a PNP transistor with another PNP transistor; the same for an NPN transistor. Also, make sure the pin-out is correct before soldering it in place and powering up.

Datasheets are extremely useful for checking the specification when looking for an alternative to an obsolete device, for identifying package type, lead arrangement, or general characteristics and usage.

Because a transistor is just two back-to-back diodes with a common junction terminal, it is easy to test if it is working.

Testing Transistors

There is a simple way of testing an unknown suspect transistor to see if it is basically sound or has failed, and to determine whether it is PNP or NPN. As a transistor is basically made up of two diodes connected together back-to-back, then all you need is a multimeter set to the resistance range or to the diode resistance setting that many multimeters now have.

Testing the "resistance" between the emitter, base, and collector will reveal if the transistor under test is of the PNP or NPN type, and if one or both of the back-to-back diodes have failed.

First, connect one of the multimeter leads to the emitter and the other to the base, and note down the multimeter reading. Now, reverse the leads and again note down the reading. Repeat the above using the base and collector, then finally with collector and emitter, always noting down the readings.

This will result in a total of six tests with two resistance readings for each test. Matching the results with the following table will reveal the type and condition of the transistor.

Become familiar with the table opposite, to speed up your transistor fault-finding skills.

Between transistor terminals:		PNP	NPN
Emitter	Base	Low Resistance	High Resistance
Base	Emitter	High Resistance	Low Resistance
Base	Collector	High Resistance	Low Resistance
Collector	Base	Low Resistance	High Resistance
Collector	Emitter	High Resistance	High Resistance
Emitter	Collector	High Resistance	High Resistance

The results are interpreted as follows:

- High/Low readings will identify if the device is PNP or NPN.
- Emitter/Base terminals – the emitter to base junction should act like a normal diode and conduct one way only.
- Base/Collector terminals – the collector to base junction should act like a normal diode and conduct one way only.
- Collector/Emitter terminals – the emitter to collector junction should not conduct in either direction.

9 Further Devices

Learn about other types of devices commonly used in electronic circuits, their construction, how they work, and the specific applications they have been designed for.

Optical Devices

Having learned the basic principles of semiconductors, it is now time to look at more advanced devices. Most of the components covered in the next few pages are a logical development of the theory detailed in the previous chapters, so their function and use should be easy to understand.

Don't forget

Single LEDs are now available that can emit more than one color depending on how they are biased.

7-segment LED display

This is simply seven LEDs grouped together in a rectangular manner as shown in the images. The seven LEDs (or segments) can be illuminated in a particular order to form numerical digits or the letters A to F, meaning that they can display decimal and hexadecimal figures.

An additional eighth LED is usually included in the package for use as a decimal point (or DP). 7-segment displays can be connected together to display groups of numbers when necessary.

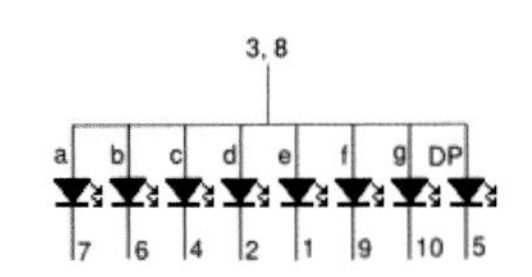

From the equivalent circuit diagram shown above of a simple display, you can see that all of the anodes (positive terminals) of the seven LED segments are connected together and brought out to pins 3 and 8, which are common. This type is called a *common anode device*.

Each one of the seven LEDs is allocated a position in the display and labeled *a* through *g*, with the decimal point LED labeled DP. Each cathode (negative terminal) is then connected to a corresponding pin on the device package. The diagram opposite shows the complete configuration of a typical 7-segment LED display device.

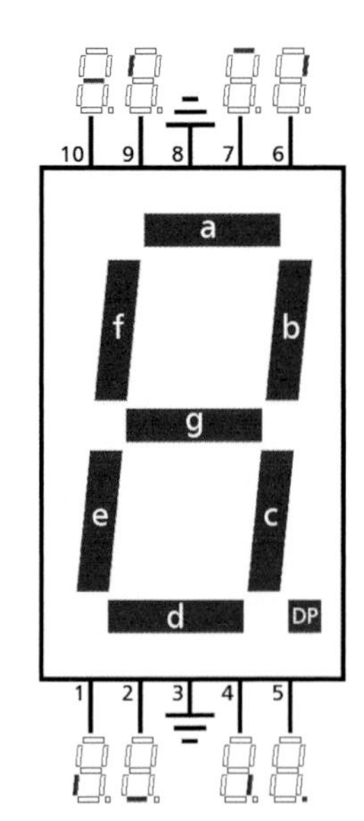

A *common cathode* version is also available where all of the cathodes are connected together and the anodes connect to individual pins. Both types are very popular as they are versatile, available in different sizes, and consume little power.

7-segment displays are available in a range of sizes and also in LCD form, not just LED.

Digit display

The table below details which segments require powering up to display the digits 0 to 9 and characters A to F. Note that some characters are displayed as lower case due to the limitations of there being only seven segments. For completeness, the device pin numbers are also shown next to the segment labels.

Display	a (7)	b (6)	c (4)	d (2)	e (1)	f (9)	g (10)
0	on	on	on	on	on	on	off
1	off	on	on	off	off	off	off
2	on	on	off	on	on	off	on
3	on	on	on	on	off	off	on
4	off	on	on	off	off	on	on
5	on	off	on	on	off	on	on
6	on	off	on	on	on	on	on
7	on	on	on	off	off	off	off
8	on	on	on	on	on	on	on
9	on	on	on	on	off	on	on
A	on	on	on	off	on	on	on
b	off	off	on	on	on	on	on
C	on	off	off	on	on	on	off
d	off	on	on	on	on	off	on
E	on	off	off	on	on	on	on
F	on	off	off	off	on	on	on

Be sure to correctly identify whether a display is common anode or common cathode to avoid possible damage to the device when it is powered up.

7-segment displays are used in many applications – on petrol pumps and in calculators, for example – so you may already be familiar with them. The LED variety is rapidly being superseded by the LCD (Liquid Crystal Display) version that uses even less power.

Bar graph rows of LEDs are also available to display red when a signal reaches a certain level, just as you would see marked on the scale of an analog meter.

Bar graph indicator

This is simply a row of LEDs in one package and is often used in place of a meter to indicate the level of a signal.

Voltage Regulators

Electronic circuits are designed to operate at a specific voltage, and although most devices will tolerate a small increase in the supply voltage, there will come a point at which something might fail due to over-voltage.

Likewise, the circuit may not function within the design specification if the supply voltage deviates too far up or down from the correct value. For this reason, any voltage source for powering an electronic circuit should be regulated to ensure voltage stability and, if necessary, current control.

Voltage regulators are manufactured for most of the common circuit supply voltages, such as 5 V, 9 V and 12 V, etc.

Electronic equipment that incorporates an internal power supply will have some form of voltage regulation employed between the output of the supply circuit and the main electronic circuit.

External power supplies or mains adapters such as a mobile phone charger will also incorporate voltage regulation circuitry once the mains AC has been reduced and rectified to DC. The stabilized DC output voltage it is designed to deliver, and current rating, will be marked somewhere on the body of the adapter.

Voltage regulation actually provides a number of features: a constant voltage is assured (within reason) even if the supply voltage to the regulator circuit fluctuates up or down; an over-current load can be protected against if required; and, most importantly, the regulated voltage will remain at the fixed level even when the load increases and more current is required.

Regulator circuit

A simple voltage regulator circuit is shown here. It is made up of discrete components, meaning that the circuit uses individual components such as diodes, resistor, transistor, etc. This circuit will provide quite adequate regulation, but with advances in manufacturing techniques it has become possible to incorporate parts of the circuit into single devices, such as a bridge rectifier instead of using four separate diodes.

A power supply would normally also include suitable capacitors either side of the regulator circuit to help with voltage smoothing.

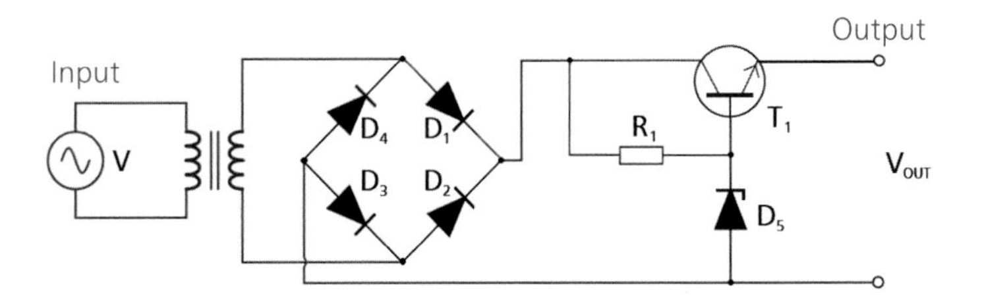

IC voltage regulator

The regulator section of the circuit shown on the previous page is available as a single device with a wide range of voltage and current ratings. The term *voltage regulator* commonly refers to an integrated circuit (IC) that provides a constant fixed output voltage regardless of a change in the load or input voltage.

A *voltage regulator* is the common term for a device containing all of the regulator circuit on a single IC chip.

Some of these regulators feature quite simple circuits, whilst giving extremely good regulation. Some are more complex and include ripple smoothing, overload protection, or variable voltage capability. They are also available for + or - output voltage.

How the voltage regulation is achieved depends on the design of the integrated circuit; the most common method being linear regulation. This form of voltage regulator uses automatic resistance adjustment via a feedback loop to compensate for changes in both load and input to keep the output voltage at a constant level.

Voltage regulator devices are commonly available in "+" or "-" voltage output form.

With the advent of device integration, the voltage regulator has become one of the most widely used devices. A typical example is the LM7812 shown opposite. Despite its small size (about 10 x 16 mm, excluding pins) this 3-pin linear voltage regulator will provide a constant +12 volt 1 amp output for an input voltage of up to 35 volts.

To give an idea of how much is packed into this small device compared with the original resistor, Zener diode, and transistor regulator circuit, the block diagram and pin arrangement from the Fairchild Semiconductor® LM78xx series datasheet is shown below.

Learn to make good use of manufacturers' datasheets, as they provide the full device specification and include suggested uses and circuits.

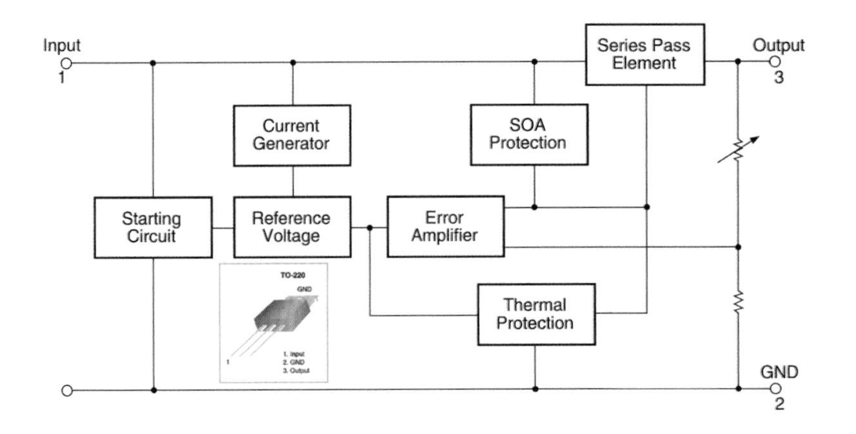

...cont'd

Main disadvantage

Popular as this type of voltage regulator is, it does have a serious drawback ultimately resulting from how it functions. Because it basically behaves like a resistor to stabilize the voltage, a lot of energy gets wasted due to the conversion of any resisted current into heat.

Ideally, linear voltage regulators should best be used where power requirement is low, and output to input voltage difference is minimal.

This drawback makes the linear voltage regulator ideally suited for use where the difference between output and input voltages is minimal and the power requirement is low. Good design of the transformer and rectifier circuit delivering the input voltage to the regulator will minimize this energy loss. However, it remains extremely popular because it is very cheap!

IC regulator circuit

The diagram below shows how the voltage regulator replaces the discrete components in the regulator circuit on page 102.

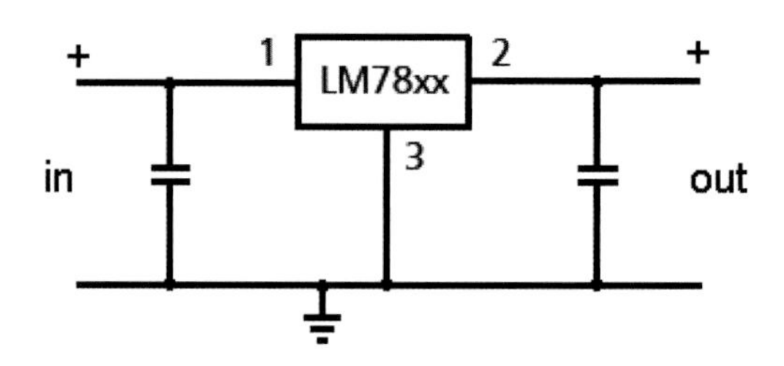

The capacitors shown on the voltage input and output are used to help smooth out any ripple.

Variable voltage regulator

It is also possible to produce a variable voltage-regulated supply with the addition of just a few components and using a slightly modified version of the fixed voltage regulator, such as the LM317, called a *variable voltage regulator*.

The LM317 is not interchangeable with the LM78xx series, as the pin arrangement is different. Always check the datasheet of a device.

This device is an adjustable 3-terminal positive voltage regulator that can supply 1.5 amps over an output range of 1.25 volts to just over 30 volts. It uses two resistors – one fixed and other variable – to set the output voltage to the desired level.

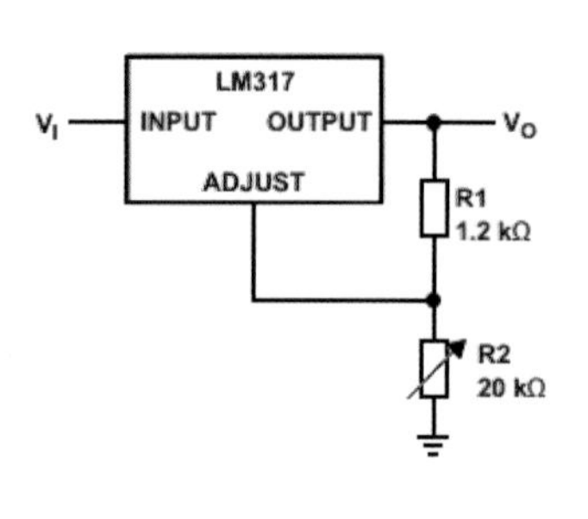

Field-Effect Transistor

The *field-effect transistor* (or *FET*) is a three-terminal unipolar semiconductor device. It has very similar characteristics to those of the bipolar transistor, such as instant operation and high efficiency, whilst being a robust and cheap device. An FET can be used to replace the equivalent bipolar junction transistor in most electronic circuit applications.

A field effect transistor is basically like a bipolar transistor but without the junction, hence the term *unipolar*.

Construction

Unlike the normal transistor, which is current-operated, the FET is a voltage-operated device that is constructed with no P-N junction within the main current-carrying path between two terminals called the *source* (S) and the *drain* (D). These terminals are the respective equivalents of the emitter and collector of the bipolar transistor. The current path between these two terminals is called the *channel*, which may be made of either a P-type or an N-type semiconductor material. A third connection called the *gate* (G) of opposite polarity material (N-type or P-type) is added near the center of the channel. This is equivalent to the base terminal.

The FET is a very efficient device and can replace most equivalent specification transistors in a circuit.

From this construction, the FET is also called the *junction field-effect transistor* or *JFET*. It works by using the electric field generated by the input voltage, which is how it gets the name of *field-effect*. The invention of the FET was to lead to the solid state "electronics revolution".

Advantages

A problem with the bipolar junction transistor is that it has a low input impedance, because the base-emitter diode is forward biased. This can cause "loading" of the signal source feeding the base of the transistor. Because an FET works with a reverse biased input junction, it has a very high input impedance and so the signal source loading is minimized.

FETs can also be made much smaller than their equivalent bipolar junction transistors. Together with their low power consumption and dissipation, this makes them ideal for use in integrated circuits such as the CMOS (Complementary Metal-Oxide-Semiconductor) range of logic chips. Another version of the FET is the MOSFET (Metal Oxide Semiconductor Field Effect Transistor).

When substituting a transistor with an FET, always check the datasheet to confirm if the biasing is suitable or whether it needs altering.

Thyristor

For many years, the thyristor was always known as a *silicon controlled rectifier* (SCR).

Originally called a *silicon controlled rectifier* or *SCR*, the *thyristor*, as it is now more commonly known, is very similar in construction to the transistor, but in operation is just like the diode in that it conducts current in one direction only.

Unlike a diode, though, the thyristor is basically a form of solid state switch. It can be made to operate as either an open-circuit switch or as a rectifying diode depending upon how it is triggered. This means that a thyristor can only ever be a switching device and cannot be used for other applications such as an amplifier.

Construction

A thyristor is a multi-layer semiconductor device that, once turned into the ON state, behaves just like a rectifying diode. It is really a four-layered P-N-P-N semiconductor rectifier; the flow of current between two electrodes being triggered by an appropriate signal at a third electrode. This means that the thyristor usually has just three electrodes – called the *anode*, the *cathode*, and the *gate* – to do the on/off triggering.

Anode (+) Cathode (–) Gate

Operation

Its operation is very simple. The thyristor conducts only when the gate receives a current trigger, and it continues conducting until the anode to cathode current falls below the minimum latching level. It can handle more power and is therefore a more efficient two-state semiconductor switch compared with a switching transistor.

The thyristor was originally called a *silicon controlled rectifier* because of its construction: *silicon* from the multi-layer semiconductor structure, *controlled* because of the way it is switched on or off using the gate, and *rectifier* because it then behaves like a rectifying diode. The circuit symbol for the thyristor suggests it is like a controlled rectifying diode.

The thyristor is basically just like a diode whose function you can control rather than it being set during manufacture.

Characteristics

- Small gate current controls a larger anode current.
- Can operate only in the switching mode.
- Acts like a rectifying diode once it is triggered on.
- Once on, conducts even when gate current is no longer applied, providing anode current is above the latching current level.

Operational Amplifiers

Also called an *op-amp*, the operational amplifier has become one of the basic building blocks of the analog electronic circuit. It is a linear device consisting, in its simplest form, of a DC-coupled high-gain electronic voltage amplifier with a differential input and a single output. The op-amp is also called a *differential amplifier*.

Op-amps are extremely versatile devices and can be used for many applications, not just as small signal amplifiers.

Op-amps actually started life many years ago in analog computers where they were used for performing mathematical operations. There are many versions of the standard IC op-amp and, as they cost very little to produce and are so versatile, they are now amongst the most widely used electronic devices and can be found in a vast array of industrial and consumer devices.

Packaging

A common operational amplifier is the LM741. It is usually available in an 8-pin dual in-line (DIL) IC package as shown below, along with the circuit symbol for an op-amp.

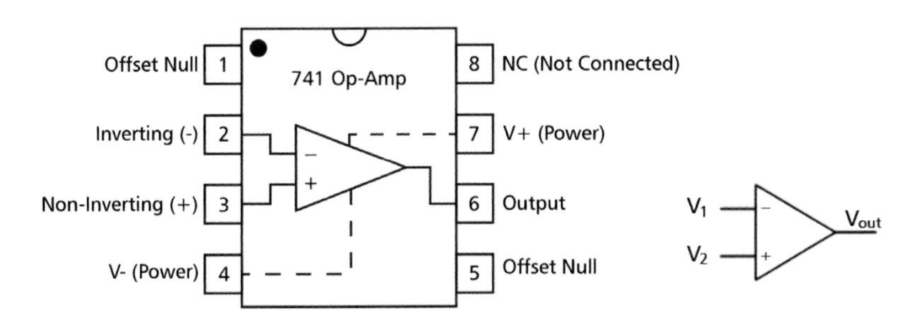

Op-amps share the same type of packaging as some digital IC devices.

16-pin packages are also available but these often contain two or even four op-amps. It is worth noting that although many of the wide variety of op-amps manufactured use the pin-outs shown above, not all do. As with transistor packaging, there is no single standard. Also note that whilst the circuit symbol given above is the one most commonly used, the power supply pins are sometimes included in some versions, not just inputs and output.

Being produced with similar specifications by a wide number of manufacturers, many op-amps are interchangeable, but always check the pin-out and voltages on the datasheets first.

Background

Earlier in this book you learned that a transistor can be used as an amplifier. However, the operational amplifier doesn't just amplify a signal. It is a much more sophisticated device whilst still being hardly any bigger than a single transistor! In simple form, the basic op-amp uses about seven transistors. The op-amp is a highly flexible integrated circuit that can be configured to perform so many common tasks.

...cont'd

The operational amplifier was actually developed many years before the integrated circuit. Early op-amps were made using valves (or tubes), and formed the basis for a mathematical circuit that could add, subtract, multiply, divide, etc. For use in analog computers, the aim was to make all of the op-amps in the circuits identical, hence standardization crept in. The op-amp became a very useful circuit for instrumentation and industrial control.

Op-amps are three-terminal devices; that is, a "+" and a "-" input and one single output.

Specification

There are many types of op-amp with similar specifications, such as the TL071, LF356, and CA3140, but for the purpose of this introduction to operational amplifiers we will continue looking at the LM741, as it is the most popular. It is made by different manufacturers and sometimes has different labels. This includes the NE741 and μA741, both of which will do the same thing.

The op-amp requires a positive and a negative power supply, typically in the region of +/-12 to +/-18 volts. The input voltage is listed as +/-15V and must always be less than the power supply voltages to avoid damage to the internal circuit. The two inputs are called *differential inputs* and consist of a non-inverting input (+) and an inverting input (–). Ideally, the op-amp amplifies only the difference in voltage between the two, which is called the *differential input voltage*.

Op-amps require a dual voltage supply: one positive and the other negative. Both should be the same level.

Fundamentally a voltage amplifying device, the op-amp is designed to be used with external feedback components (resistors and capacitors) between its output and input terminals. It is these feedback components that determine how the amplifier will function or operate, hence the name *operational amplifier*. It is by virtue of the different resistance/capacitance configurations that the amplifier can perform a wide variety of different operations.

Some applications

- In oscillators, waveform generators, and differential amplifiers.
- In analog-to-digital converters (ADC) and digital-to-analog converters (DAC).
- Video frequency and audio frequency pre-amplifiers/buffers.
- In voltage and current regulators and for voltage clamping.
- As filters in signal processing circuits.

Crystals

Whenever there is a need for a very stable frequency source you are bound to find a crystal somewhere in the circuit. Crystals, also called *xtals* or, more accurately, *quartz crystals*, are extensively used for providing a constant frequency output under varying load conditions such as variations in temperature or the load on the circuit, as well as changes to its DC power supply voltage.

Virtually every appliance we use today, from a small mobile telephone to a television set, uses one or more quartz crystals.

Operation

A quartz crystal exhibits a property discovered in 1880 and known as the *piezoelectric effect*. Simply explained, when mechanical pressure is applied across the faces of a thin slice of crystal, a voltage that is proportional to that mechanical pressure appears across the crystal. The ignition system on gas fires used to work using this effect to generate a spark to light the gas.

The opposite is also true; applying a voltage across a crystal causes distortion in the crystal. Applying an alternate voltage causes the crystal to vibrate at its natural frequency, and it is this property that makes the crystal ideal for accurately controlling the frequency of an oscillator.

Crystal oscillator circuits utilize this inverse piezoelectric effect. In the crystal, the applied electric field produces a mechanical deformation for generating an electrical signal of a particular frequency. The crystal used in a crystal oscillator is a very small, thin wafer of cut quartz with the two parallel surfaces metalized to make two electrical connections. It is the size and thickness of the quartz wafer that determines the frequency at which the crystal will oscillate, also called the *fundamental frequency*. Once cut and shaped, the crystal cannot be used at any other frequency.

Being so accurate, crystals are extensively used for timekeeping. A watch or clock, whether analog or digital, will certainly contain one.

Frequency

Crystals can be manufactured for oscillation over a wide range of frequencies, from a few kilohertz up to several hundred megahertz.

Crystal oscillator types

- Temperature-compensated crystal oscillator (TCXO).
- Voltage-controlled crystal oscillator (VCXO).
- Oven-controlled crystal oscillator (OCXO).
- Temperature-sensing crystal oscillator, an adaptation of the TCXO (TSXO).

Loudspeakers

The job of a loudspeaker is to convert an electrical signal into sound. Although some speakers are designed to handle a wide range of frequencies, they could not possibly be efficient across the whole range and reproduce accurate quality sound.

All loudspeakers, no matter what type, work by moving air. This movement is detected by the human ear as sound.

Types of speakers

Some speakers are therefore designed to handle only a limited range of frequencies. Common types are listed below:

- **Full-range** – designed to cover the entire audio frequency range. Typically 3 to 8 inches (7.6 to 20.3 cm) in diameter to permit reasonable high-frequency response.
- **Woofer** – designed to reproduce only low frequencies. Also called a *bass speaker*. Powerful and large in size to aid faithful bass sound reproduction.
- **Mid-range** – designed to reproduce a band of frequencies generally between 1-6 kHz, otherwise known as the "mid" frequencies (between the woofer and tweeter range).
- **Tweeter** – a high-frequency speaker that reproduces the highest frequencies in the audio range.

For high-quality sound, a tweeter, mid-range, and woofer are often assembled together in one housing.

Construction

As with most devices, there are several different technologies and approaches used to manufacture loudspeakers. The moving coil type of loudspeaker is the most common, and consists of a cone attached to a coil held suspended in a magnetic field.

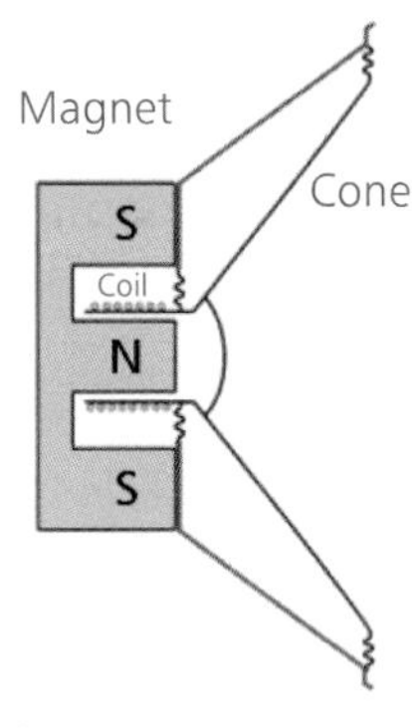

When the current flowing through the coil varies as a resulting of the electrical audio signal, the magnetic force generated causes the coil, and hence the cone and the air around it, to move. This results in the loudspeaker converting the electrical audio signal into sound.

The familiar circuit symbol for a loudspeaker is shown here. A microphone works in a very similar (but reverse) way to the loudspeaker.

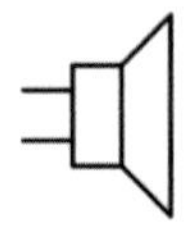

10 Construction Methods

This chapter shows you how an electronic circuit is built, from initial design through to construction. The use of simulation software is introduced, and you are shown how a PCB is made.

Overview

There are a number of ways to design and then build an electronic circuit. Now, that you're familiar with electronics and the various components, then the simplest method is to first roughly draw your circuit and then collect together the components, connect them together and see if it works. This is rather hit and miss, though many constructors still use this method for simple circuits.

There are far better ways of designing and building your circuit. One of the obvious problems with the simple method outlined above is that although you hope that you have calculated or selected the correct components, your circuit may not work properly (or at all!) when everything is connected together and you apply power.

The easiest way to design and construct a circuit is to use a computer and a simulation package. Not only will the software help with the design and testing, but it will usually also produce a list of components and a track layout for creating a printed circuit board (PCB).

Now, you have to find why the circuit doesn't work. Are the resistor or capacitor values correct? Have suitable transistors and diodes been selected? Has the circuit been wired up correctly? Etc. There are a whole host of reasons why, and there isn't just one overall solution to cover everything.

Each problem will have to be looked at separately so, to help make things a bit easier, the process of circuit design and construction is broken down into different stages. Work can be checked at the completion of each stage, and only when you are satisfied that everything is in order and no mistakes have been made can you then progress to the next stage.

Fortunately, with advances in technology, there are a number of aids that you can use to help you complete some of these stages. For example, computer programs are readily available that will greatly assist you at the design stage. Most contain a library of basic circuits that you can integrate into your circuit, making your job easier, as you may be able to take shortcuts in the design.

Other inbuilt features include simulating your circuit before you move on, to building it so that any errors are flagged at an early stage. These can be corrected, simulation run again, and so on until your circuit is working properly – in principle, at least.

Finally, you are ready to build your circuit, but what if you make a mistake when connecting everything together? Again, there are a number of options as you will see. The simulation software may have provided a components layout and connection diagram, but this doesn't mean to say you managed to follow it exactly right.

Simulation Packages

Because there are so many things to consider when designing electronic circuits, using a simulation package to develop and test your design is a really good idea. There are a number of circuit simulators available; some you can download free of charge from the internet, whilst more sophisticated ones you have to pay for. Examples are illustrated on pages 115-118.

Simulation software can find a mistake in your circuit design and allow you to quickly correct it.

Using simulation software

The use of simulation programs or similar type of software has become increasingly popular. Even if you haven't designed a circuit yourself, but have seen an interesting one in a book or magazine that you wish to build, it is still a good idea to copy that circuit into the simulation program so that you can see how it should perform, plus the software can provide you with construction templates, a wiring diagram, and so on.

The simulation software package you use should have a library containing the symbols and function of all the common electronic components such as resistor, capacitor, diode, voltage rectifier, transistor, etc. Updated libraries can often be downloaded from the internet and added to the simulation package.

Should you really need to, you can usually add any device not included to the library yourself, together with the component's specification for accurate simulation. In the case of transistors and op-amps it should be possible to select an existing device from the library, edit its specification and pin arrangement (if necessary) to that for the device you wish to add, then save these details under a new device name.

Simulation software makes it easy to change component values or re-arrange a circuit.

Creating the circuit

To create a circuit using simulation software, just do the following steps, which are common to most simulation packages:

1. Select the first device from the library, such as a resistor, and place it in the drawing area
2. Allocate it a label, such as R1
3. Allocate it a value, such as 1.5 kΩ
4. Repeat Steps 1-3 until all components have been placed

...cont'd

Make sure to allocate the correct values to all of the resistors and capacitors as they are placed in position.

5. From the library, select a suitable power source and place it on the drawing area

6. Select a suitable size of connecting line and carefully connect all of the components on the drawing area together as per the circuit diagram

Components that are regularly used in circuits, such as some transistors and diodes, may already be pre-programmed with their working characteristics, but you will still need to give each one a unique label as you add it to the circuit, such as Q1 or D1.

The characteristics of many commonly used devices will be pre-programmed in libraries in the software package.

An example of how a circuit will look when laid out using typical simulation software is shown here:

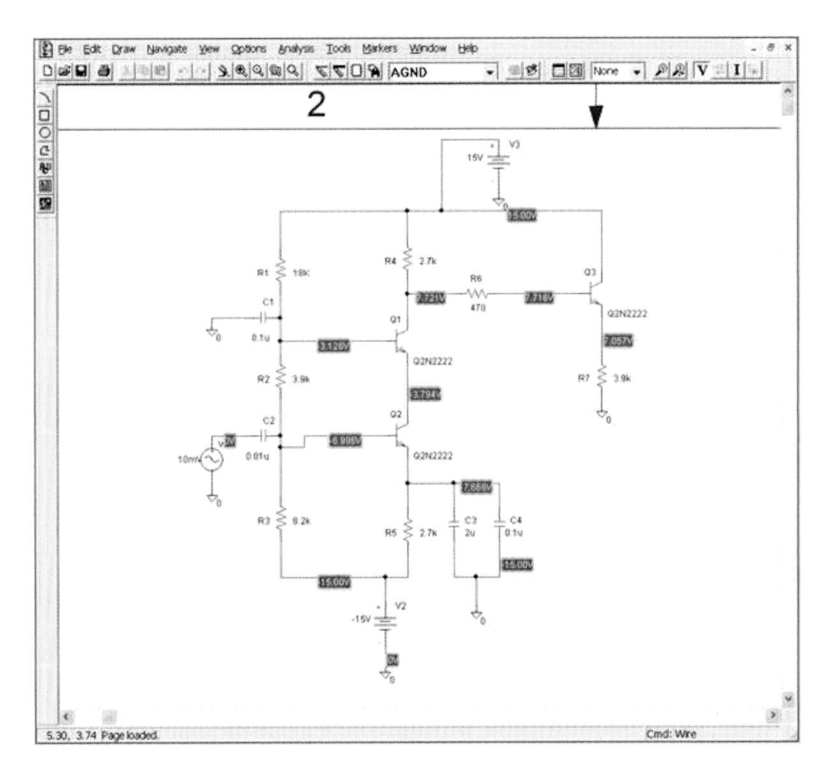

For simplicity, most packages use the alternative resistor style and basic transistor symbols. This screenshot contains seven resistors (R1–R7), four capacitors (C1–C4), three transistors (Q1–Q3), two DC power sources (V2, V3), and one AC signal input. A virtual probe has been used to display voltage at various points around the circuit so comparison can be made against calculated values.

SPICE Software

This widely used program is a general-purpose electronic circuit simulator that is suitable for both analog and digital circuits. SPICE, which stands for Simulation Program with Integrated Circuit Emphasis, is incorporated in a number of derivatives such as LTspice, Pspice, and Xspice. They are very similar in operation, but have been developed for better handling of specific applications.

Many electronic circuit design and simulation packages are based on the SPICE software.

The main function is to check the integrity of circuit designs and to predict how a circuit will operate or function. It can cope with some very complex designs, and can be used for testing how a circuit will behave when different components are altered. This means that a circuit can be tested to see how well within specification it is, how it behaves when components are changed, and what might happen if the input signal is too large or small.

Test instruments

A range of virtual test instruments is available, so voltage and current measurements can be made for comparing against expected results. Waveforms can even be observed using a virtual oscilloscope to provide realistic run-time traces similar to the example shown below.

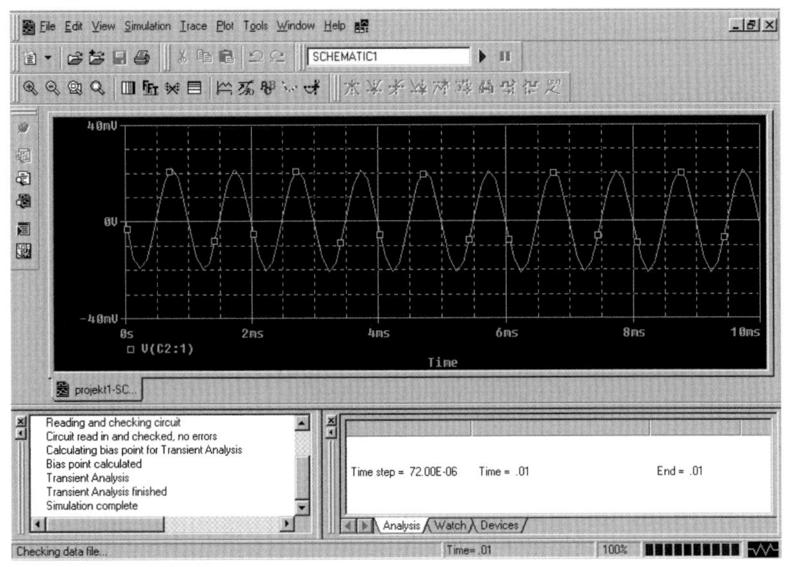

Remember that this is all simulation; what you see is a prediction of how the circuit should function in ideal circumstances prior to actually building it. However, you should always bear in mind that actual circuit performance may be slightly different, due to variations in physical component values due to tolerance.

Make good use of the virtual test instruments that come with the software to ensure your circuit is working exactly as intended.

Multisim

Many packages come equipped with a library of sample circuits.

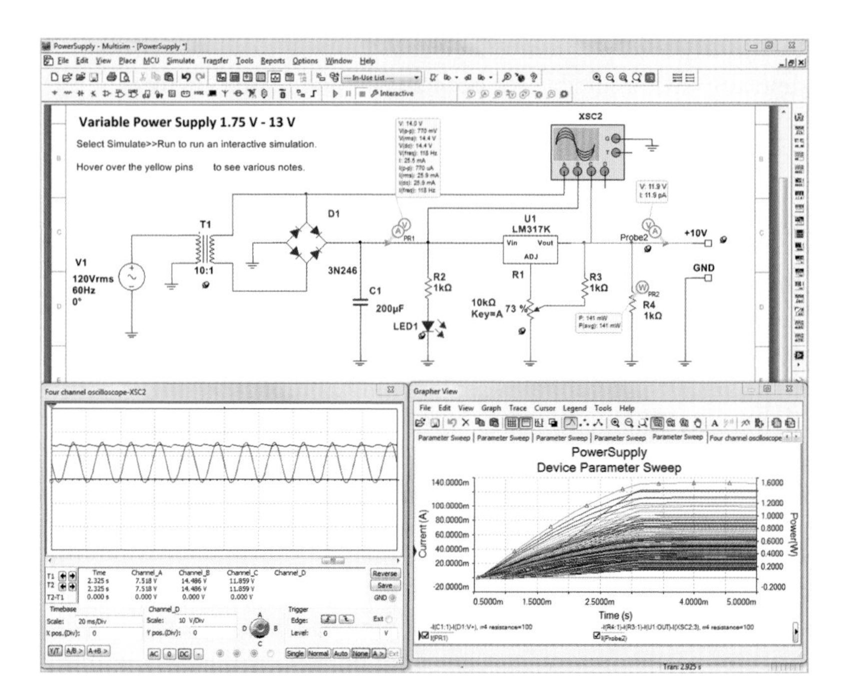

Use a sample circuit to quickly teach you how to use the software.

Shown above is another well-known powerful simulation package called *Multisim* from National Instruments™. Also known as *NI Multisim*, this software is part of a suite of circuit design programs that includes NI Ultiboard for designing the printed circuit board layout. It actually employs the original SPICE-based software, and is well supported and widely used in education and industry.

This example from the NI Multisim tutorial readily demonstrates many of the topics we have looked at so far. It shows a variable output voltage power supply circuit utilizing bridge rectification and a version of the LM317 linear voltage regulator mentioned in Chapter 9. The output voltage is adjusted by the variable resistor labeled R1 below the regulator IC.

The software shows you how the circuit works under ideal conditions, but component tolerances might alter this slightly when the circuit is actually built.

Voltage and current probes are also employed, together with a virtual four-channel oscilloscope (XSC2). Three scope traces show the sine wave output from transformer T1 in red, the rectified voltage leaving the bridge in blue (with some ripple), and the fully smoothed and stable DC output in green. This demonstrates how well the circuit should ideally work.

Multisim is not free, though a trial version can be downloaded. However, there are many good free online products also available.

Free Programs

Downloadable software

Fortunately there are a number of electronic circuit simulators available under a free software license. A quick search on the internet will produce a choice of free packages that provide all of the functions and facilities you will need to get started with electronic circuit design without having to spend money. Many have versions that will run on different computer operating systems. The following are just a couple of downloadable examples.

Many packages are available for running on different computer operating systems.

Proteus

This is SPICE-based software that runs on the Windows operating system. The trial version can be downloaded for free, although it does have certain limitations in that you cannot save your work or print out circuit diagrams or layouts. It is, however, not time-limited, and contains an extensive set of sample designs to help you to become familiar with simulation software for free. Tutorials are also available for download in PDF format.

Some software is available as a free trial for a short period so you can evaluate it.

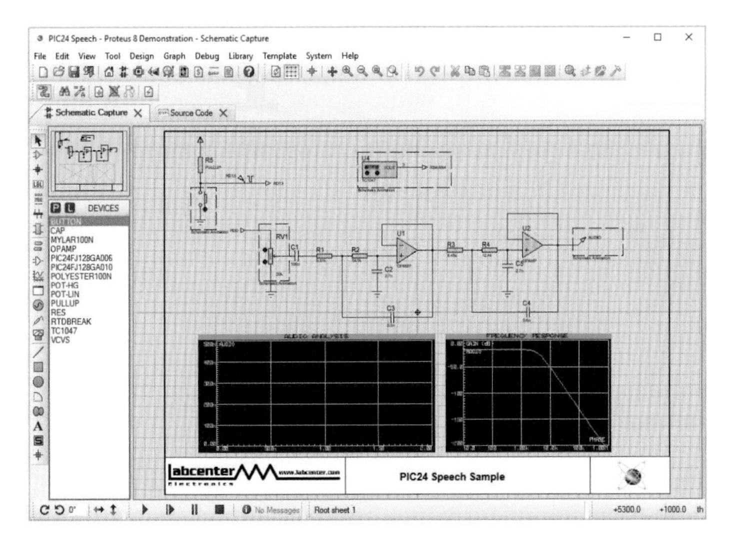

LTspice

Being freeware, LTspice can be downloaded and used without payment. As the name suggests, it also implements SPICE electronic circuit simulation and is the most widely distributed and used SPICE program in the industry. It is produced by semiconductor manufacturer Linear Technology (hence LT), and so includes macro models for most of their regulators, op-amps, transistors, and passive components. Not designed specifically for logic circuits, LTspice does support simple logic gate simulation.

Although versatile, not all packages are suitable for use in designing digital circuits.

...cont'd

Online software

If you don't want to download and install simulation software on your computer then there is also online software you can use. This is probably the best way for you to be introduced to circuit simulation, provided you have a good internet connection, as all work is done online. Again, here are some popular examples.

Online software is ideal for new users as there is no installation or setup – you are ready to go as soon as you log in.

PartSim

PartSim is free and easy to use. It runs in your web browser and includes a full SPICE simulation engine, web-based schematic capture tool, and a graphical waveform viewer. To access PartSim all you need to do is search for it on the internet – it immediately loads up once you click on the web address.

It is entirely free to use and includes a large number of parts and components from different manufacturers, making PartSim a practical electronic circuit design simulator and good for beginners. The downside is that it is not as powerful as some simulators plus the library contains lots of op-amps but few other ICs.

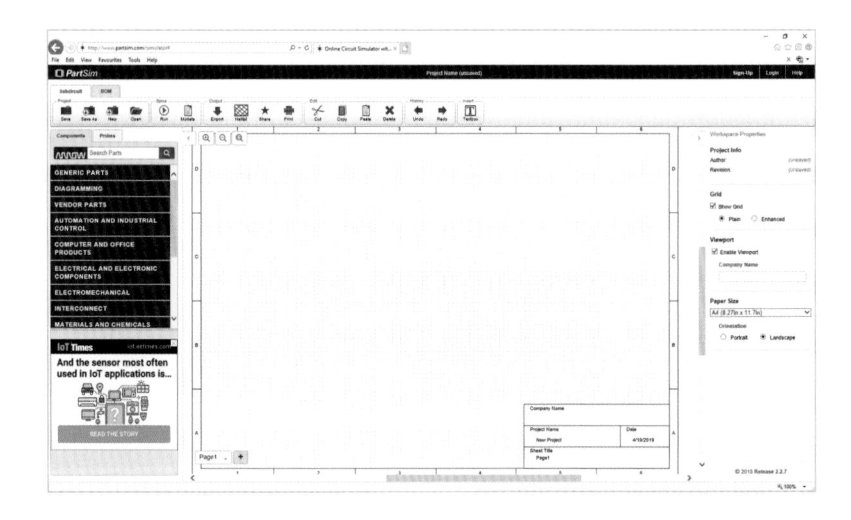

Always save your design at regular intervals in case of unexpected computer failure or internet access drop-out.

EasyEDA

A free and easy-to-use circuit simulator and PCB layout package, EasyEDA is very popular and features a large parts library.

Circuit Sims

An extremely basic circuit simulator that is perfectly suited to understanding simple circuits and electronics. Runs on any browser.

Circuit Construction

Whatever method you use to design a circuit, or perhaps you have found a circuit you would like to make, you will eventually be ready to actually construct it. There are a number of different ways of building electronic circuits. Which one you decide to use depends on a number of things.

Breadboard construction is ideally suited to small circuits with relatively few components.

Circuit construction methods

If you are satisfied that the design is complete, you will probably want to build it using a printed circuit board, or PCB. However, if you think you will need to do further testing and possibly change some of the components, then you should use a fairly simple construction method that allows you to easily change the circuit if it doesn't work properly.

There are three construction methods you can use; these are detailed in the following table:

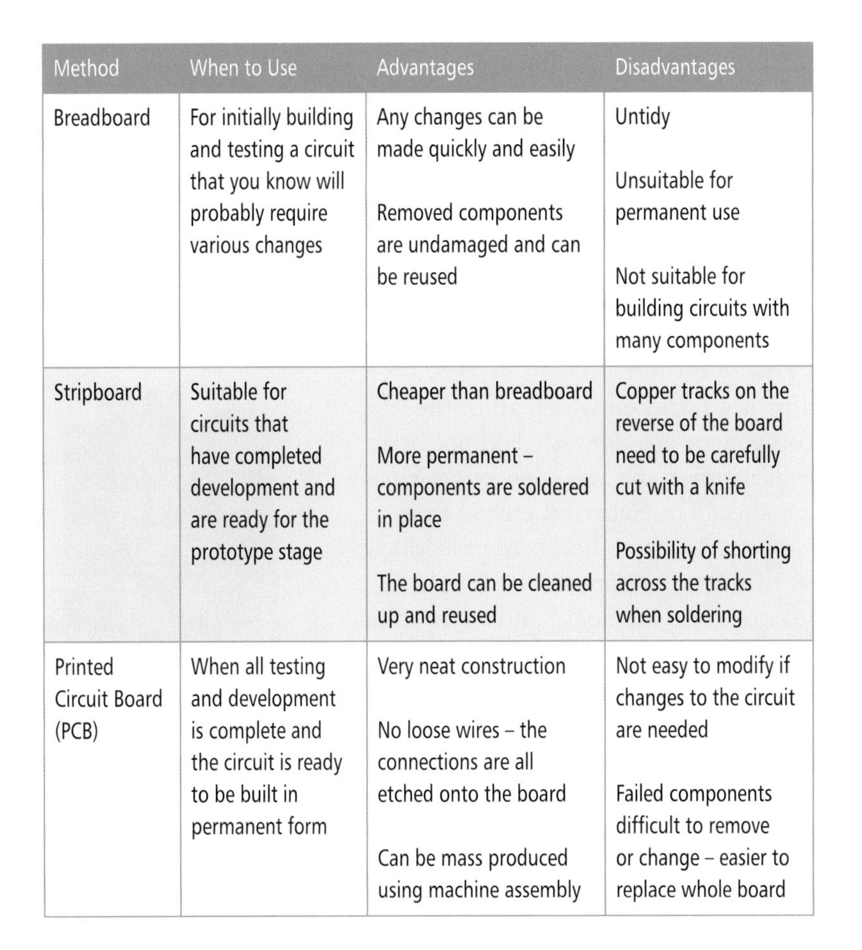

Method	When to Use	Advantages	Disadvantages
Breadboard	For initially building and testing a circuit that you know will probably require various changes	Any changes can be made quickly and easily Removed components are undamaged and can be reused	Untidy Unsuitable for permanent use Not suitable for building circuits with many components
Stripboard	Suitable for circuits that have completed development and are ready for the prototype stage	Cheaper than breadboard More permanent – components are soldered in place The board can be cleaned up and reused	Copper tracks on the reverse of the board need to be carefully cut with a knife Possibility of shorting across the tracks when soldering
Printed Circuit Board (PCB)	When all testing and development is complete and the circuit is ready to be built in permanent form	Very neat construction No loose wires – the connections are all etched onto the board Can be mass produced using machine assembly	Not easy to modify if changes to the circuit are needed Failed components difficult to remove or change – easier to replace whole board

Stripboard can also be used for permanent circuit construction provided it is carefully built and soldered.

Mistakes are more difficult to rectify with PCB construction.

...cont'd

Hot tip

Components used for breadboard construction can be reused as there is no soldering involved.

Breadboard

A method often used in the early stages of circuit design. Despite looking untidy, the components and wiring can be easily changed if the circuit needs modifying. It is also easy to access the various parts of the circuit with test probes. Because of its rough build, breadboard construction is not suitable for permanent use or for large circuits containing a lot of components.

Stripboard

This is a generic name for a widely-used type of electronics prototyping board. It consists of a regular grid of 0.1 inches (2.54 mm) holes with parallel strips of copper tracks running in one direction all the way across one side of the board. It is also known as Veroboard. To use the board, breaks are made in the tracks near the holes to create the equivalent of strips of wires for connecting the components together. Stripboard is not designed for use with surface-mount components.

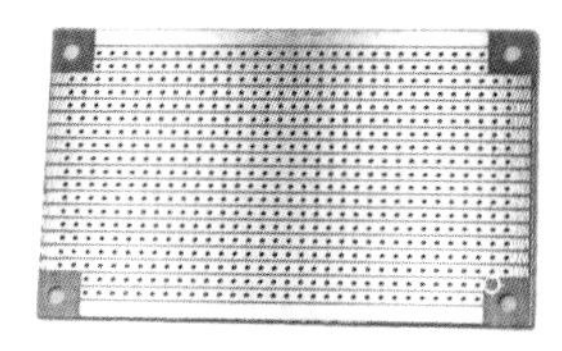

Printed circuit board (PCB)

PCB construction is usually preferred, as it produces a professional-looking result every time.

This is a board on which all of the component positions are laid out in a neat manner, and the connecting wires are already in place and etched from copper. Mounting holes are pre-drilled so that the components are simply inserted into the board and soldered in place, often automatically. The board may have tracks or connections on one side only, on either side or even built into the board on different layers. It is a sturdy form of construction but very difficult to repair or modify. It has become more economical to replace the whole PCB than attempt a repair.

Making a PCB

Although breadboard and stripboard construction make circuit design and testing simpler, they are not really suitable for permanent circuit construction. Stripboard could be used for circuits with very few components, but a printed circuit board would be far more suitable. The PCB makes things a lot easier by eliminating connecting wires, reducing the risk of making wiring mistakes and, if carefully designed with regard to component placement, will reduce the size of the final physical circuit.

PCB construction reduces the number of loose wires and makes the end product more robust.

Overview

A PCB is a board with all of the circuit wiring represented by copper tracks and soldering pads. These features are etched from copper cladding on one or both sides of the board to reduce the wiring, thereby reducing the faults arising due to loose connections. Once the board has been etched and drilled, the components are simply located on the board and soldered in place.

A board with only one copper-clad side is called *single-sided*; a board with both sides clad with copper is called *double-sided*. Professionally produced board can be multilayer, meaning that there are also layers of tracks buried in the center of the board. Holes on double-sided board are often "plated through" so that a track on one side of the board is connected to a track on the other side of the board.

Copper-clad board, both single and double-sided, is available in a number of sizes and quantities.

Multilayer and plated-through holes are beyond the scope of home construction but it is perfectly possible for you to create a single or double-sided PCB. Through-the-board connection is possible by simply passing a wire through a hole and soldering either side to the appropriate tracks.

Methods

Once you have completed your circuit design and tested it to ensure it is working exactly as intended, you will most likely want to produce a neater and more permanent version by using a printed circuit board.

There are three basic methods for making a PCB.

- Draw the tracks on the board by hand.
- Iron on the tracks using a template "negative".
- Use photo-etch board and ultraviolet light.

Drawing the tracks by hand can be easy and simple, but the iron-on and photo-etch board methods produce a neater result. See page 122 for further details.

...cont'd

Whichever method is chosen, it is the initial task of marking out the tracks on the board that differs; the etching process is common to all methods. Here is how to make a hand-drawn single-sided PCB.

Use an etch resist pen to draw circuits on plain copper laminate. These pens contain permanent ink with etch resist properties that is unaffected by the ferric chloride solution. Before etch resist pens were available, paint applied with a thin brush was the usual way of drawing tracks.

Drawing the tracks by hand

This is the simplest method, but good results are possible provided you are careful and neat. Before starting to make the board you need to work out on a sheet of paper that is the size of the board where you would like to position the components, and then draw in the tracks that will join everything up. Remember that the components are mounted on the side of the board that has no copper, so when you draw in the tracks for the copper side you will be looking at the components from the underneath – very important, this is the most common first attempt mistake!

It does involve a bit of trial and error – you may have to reposition components and redraw the tracks, but with a bit of patience it will be possible to achieve a result. If any of the tracks need to cross, you can always use a small wire link to join them up at the component soldering stage, but try to use as few links as possible.

If you used simulation software to design your circuit then the program may have the facility to also provide a track layout for you. If not, it's pencil and paper time. Here are the basic steps:

Always make sure that you have correctly laid out the component positions and tracks before committing the design to the board, as mistakes will be difficult to rectify.

1. Mark out the size of the board on a piece of paper. Remember you will now work from the copper side
2. With a pencil, draw on the paper where each component is to fit through the board, making sure that the positioning of the leads and any label is clearly marked
3. Now, draw in the tracks from component to component
4. When completed, carefully fix the paper in place on the copper side of the board
5. With a scribe or small nail, lightly punch through the paper onto the board where each component lead is going to pass through the board. This is to mark where the board will later be drilled for component mounting

6. Remove the paper from the board, then clean the copper side of the board with a solvent to remove any marks

7. Using an etch resist pen, carefully fill in a small pad around each component mounting position marked on the copper in Step 5. This is to aid soldering later

8. Again, using the etch resist pen, carefully hand draw the tracks on the copper. If you make a mistake, simply wipe off the track with solvent and draw it again

9. When finished, your board is ready for etching

When laying out components, don't be tempted to squeeze everything into a small space. Drawing the tracks may then be difficult and some components, such as inductors, may cause interference if mounted too close together.

Etch resist pens are cheap and easy to use but there are other methods you can use to mark the tracks on the copper. Model paint can successfully be used if put on with a fine brush. A very professional finish can also be achieved using rub-on dry transfers, sheets of which are available specifically for use on copper board. The tracks on the board shown on this page were made using old Letraset sheets found in a drawer!

Etching the board

Your printed circuit board started life as a board with one side covered in copper. The tracks you drew is where you want to keep the copper to act as connecting wires; the exposed copper is not required and is ready to be removed by a process called *etching*.

To do this, a solution of *ferric chloride* (or etching solution) is used. The board is placed in a container copper side up, and etching solution poured in until the board is well covered. The exposed copper is slowly dissolved away leaving just the tracks you drew earlier. When etching is complete, the board is taken out, and washed and cleaned to remove whatever was used to make the tracks. The end result should look like the board shown.

Take care when handling etch solution. Always clean up spills and avoid contact with eyes or exposed skin.

Populating the PCB

With the board etched, it is now ready for the components.

Don't drill component mounting holes too big – they only need to be large enough for the wires to pass through.

Drilling the PCB

Before the components can be added, the mounting holes have to be drilled. This can be done using a small drill mounted on a stand but, with a bit of care, it is also possible to drill the holes using a small handheld drill and a steady hand. Hobby drills are ideal for drilling the holes in a PCB.

The size of the drill bit used is quite small, and little bigger than the thickness of the component leads, so be careful not to use too much pressure. A larger drill bit can be used for board mounting holes or for where components require fixing to the board. This is the time to make sure all of the required holes are drilled, as it will become difficult to drill any holes you may have missed once you have started mounting the components.

Use a thin piece of tape to hold a component in place whilst turning the board over to solder it. (See pages 144-145 for more information about soldering.)

Mounting the components

With the board fully drilled, you can start adding the components. It is common to add the transistors and similar devices first, making sure the leads are fitted the correct way round. Likewise for diodes and other polarized components such as electrolytic capacitors. Normal capacitors and resistors can now be added, together with the remaining components.

Components are placed on the opposite side of the board to the tracks. Make sure the leads are straight or, if necessary, hold the component over its position and bend the leads with a pair of thin-nosed pliers so they will fit nicely through the holes. On the track side, solder the leads to the tracks, taking care not to use too much solder in case tracks become shorted. Once soldered, cut off the excess length of lead and repeat with the next component.

Solder one component at a time, and use only as much heat and solder as is needed. Too much heat could damage the component, the track itself, or even the board.

With care, the method of drawing the tracks by hand, whether with a pen, paint, or dry transfer, can produce very good results.

11 Power Supplies

Learn all about using block diagrams, the different types of power supplies available, and homemade supplies.

Block Diagrams

It isn't always necessary to study the full circuit diagram when working out how a circuit works. A method called a *block diagram* can be used. Because an electronic circuit usually consists of a number of different *stages*, each performing their own function, the complete circuit can be represented in a simpler manner by using a block to represent each stage.

It helps to first start with a block diagram when designing a new circuit.

In this way, the block diagram shows how the entire circuit functions without having to get down to the individual component level – this is only necessary when examining one particular stage. Breaking down the circuit into stages also makes fault-finding easier. For complex circuits it also means that the various stages can be designed and tested independently before they are all connected together.

Typical example

A block diagram is also a good way to start when designing a circuit. You work out the stages you need, draw out the block diagram, then tackle the actual circuit design of each stage, test it and redesign if necessary until it is working as intended, then you move on to the next stage.

Here is the block diagram for an FM radio receiver with the different stages listed underneath. As you can see, it tells you all you need to know about how the circuit will function – from a signal received by the antenna, to sound out of the loudspeaker.

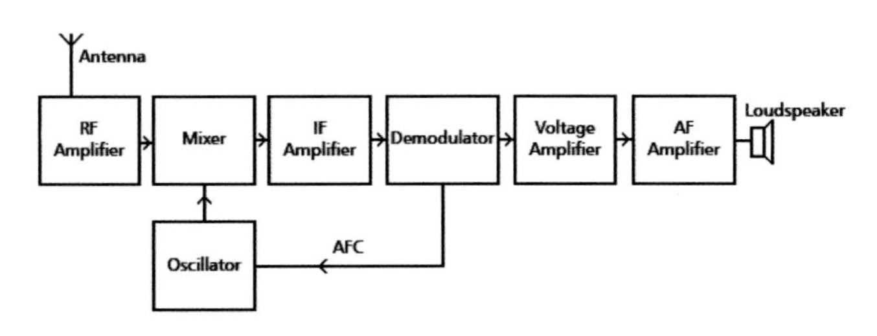

A block diagram helps to quickly identify the different stages of a circuit. These can then be independently designed and tested before putting the whole circuit together.

- RF Amplifier – amplifies the received radio signal.
- Oscillator – generates the signal for tuning the radio.
- Mixer – extracts the desired signal by mixing two frequencies.
- IF Amplifier – amplifies the intermediate signal from the mixer.
- Demodulator – extracts the audio from the signal.

- Voltage Amplifier – boosts the audio signal.
- AF Amplifier – amplifies the audio to drive a loudspeaker.
- AGC – stands for Automatic Gain Control. This function compensates for the reception of both loud and weak signals so signal levels within the circuit are kept constant, otherwise stages could become overloaded and not work within their design specification.

A power supply can be designed as an integral part of a circuit and housed inside the device or supplied later as a separate external unit.

What's missing?

Now you know how an FM radio receiver works without ever needing to look at a circuit diagram. Remove the demodulator and you would have an AM receiver. There are plenty of designs for each stage in textbooks and online, so you don't even need to start designing the circuits from scratch yourself. These are called *building blocks* – the tried and tested basic designs. But can you spot that something is missing?

The power supply

Yes, it's the power supply. Because you know that each stage will require electrical energy it is not strictly necessary to include the power supply in the block diagram.

However, power will be needed for each stage, so a power supply will have to be designed and built along the lines of the regulated power supply, circuit, and components discussed in Chapter 9, or the circuit simulation example of a variable regulated supply shown on page 116.

Power supplies can be fixed voltage or variable. Standard fixed voltages are 5 V, 9 V, and 12 V.

External power supply

It is not always necessary to incorporate the power supply into the actual circuit. Many domestic devices, for example, are supplied with a separate power unit that plugs into the mains supply and connects externally to the device. Cordless phones, internet wireless boxes, and computer games consoles are supplied like this.

Having an external power supply has two main advantages:

- The actual device can be made quite small and lightweight as it doesn't have to house the power unit.
- The manufacturer does not have to design the power supply. A ready-made one can be supplied instead, thus reducing cost.

Bench Power Supply

To save having to design and build a power supply before you can test a circuit, ready-made power supplies are available for general purpose use. These are called *bench power supplies* (or just *bench supplies*) and are ideal for general circuit design, testing, and fault finding, be it on your own circuit or a device you wish to repair.

Because different circuits often have different supply voltage and current requirements, the variable supply is the most useful. This allows you to set the voltage required by a particular circuit or device. Once set, the supply will maintain this voltage irrespective of any change in the load.

A variable voltage bench power supply is a very useful item of equipment, as it can be used to not only power complete appliances but also to power circuits that have been isolated for testing.

Current limiting

If the supply has a current-limiting facility, this is even better. You can then initially set a low current limit so that if there is a fault with the device or circuit you are testing, the amount of current it tries to consume is limited, and risk of further circuit failure is greatly reduced.

Likewise, you may have made a wiring mistake with a circuit you have built. The last thing you want is to put a short on the supply, or watch as smoke starts coming from an incorrectly wired component when you first apply the power. Current limiting will protect against this and protect your pocket from additional cost!

Two popular variable bench power supplies are shown here. They are virtually identical except for the displays; one being analog and the other digital.

Always set a low current limit when testing a new or faulty circuit. It can always be increased later when you are sure that nothing is shorting or possibly overheating.

This type of supply has very handy features. Two fixed 500 mA outputs are provided for the most common requirement of 12 V and 5 V; the latter for digital circuits. The variable output of 0-30 V is via the large red and black terminals at the lower left of the unit. Maximum current is 2.5A.

High Current Supplies

So far we have considered low current power supplies required to deliver up to 3 A or less. These are suitable for many applications but there is a lot of equipment that requires considerably more current – possibly 20 to 30 amps or sometimes more.

High current supplies are normally available with 20 A, 30 A and 50 A nominal rating (plus an additional 5 A maximum safety margin).

For example, radio transceivers (a transmitter and receiver in one unit), as used by radio hams, usually have a high current demand. The more power a supply is designed to deliver, the larger the transformer required and, consequently, the heavier the unit. Include a very large heatsink and a fan to keep it cool, and the power supply becomes even heavier.

Unless designed for low power, most transceivers are capable of delivering 100 W of output power to the antenna. A few can deliver 200 or even 400 W, and all types will be equipped with a good-sized heatsink and cooling fan or fans. This means that a transceiver is quite a weighty device so, for convenience, most are designed without an internal power supply.

Always observe caution when working with any high voltage or high current power supplies or associated circuitry. For extra safety, ensure all large capacitors are fully discharged.

The standard transceiver is designed to operate from a 13.8 V supply and with a current consumption of 22 A for the 100 W version. It would be a simple matter of taking the design for a low current power supply and uprating the specification of all the components – transformer, rectifiers, capacitors, power transistors, voltage regulator, etc.

A high-power supply built in this way would work, but there would be little protection for the device it was powering if a fault suddenly developed that caused the regulation to fail and the output voltage to rise. Transceivers are expensive devices so they need protecting. The low-current power supply designs so far considered now need sturdier components and good protection circuitry to bring them up to high-current specification.

The high-power supply shown opposite can deliver up to 25 A and also has two 6 A outputs and a 10 A output to run additional devices. It is fully overload protected and will shut down if a fault causes the voltage stabilization to fail, thus protecting the transceiver.

Most high-current external supplies are designed with one main output and one or more lower-current terminals.

Switch Mode Power Supply

Switch mode power supplies do radiate electrical noise due to the nature of how they work. This could interfere with communications or radio equipment or television reception.

Power supplies belong to two types: linear or switching. Up to now we have only looked at linear supplies, which can be large, heavy, and expensive. In comparison, *switch mode power supplies*, as they are called, are considerably smaller, lighter, and cheaper. Both types have their advantages and disadvantages but are equally capable of good performance and high current output.

Design

From the block diagram opposite, you can see that a switch mode supply is more complicated than a linear supply.

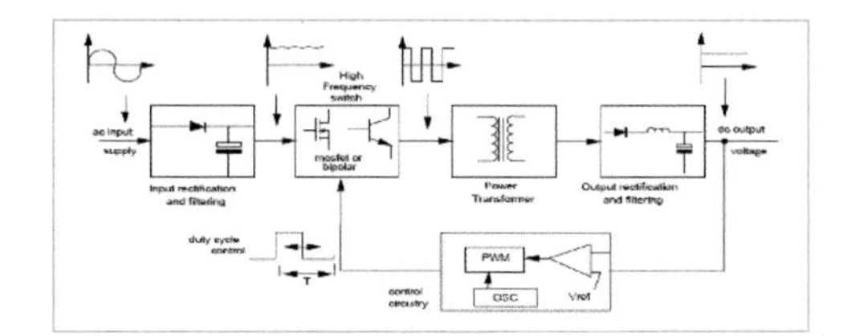

A switch mode power supply is a power converter. It utilizes specialist switching devices, such as MOSFETs (see page 105), that continuously turn on and off at high frequency. Energy is stored in the capacitors and inductors for supply power during the non-conduction state of the switching devices.

Switching supplies have a higher efficiency (up to 90%) than linear supplies, whilst being small in size. They are widely used in computers and other sensitive electronic equipment.

The main components of a switch mode power supply (SMPS) are:

- Input rectifier and filter.
- The inverter, consisting of a high frequency signal and switching devices.
- The power transformer.
- An output rectifier.
- Feedback system and circuit control.

Switching power supplies are considerably more efficient than linear power supplies.

Rectified but unregulated DC is fed to the inverter section containing fast switching electronic devices and bipolar transistors that are driven on and off. This causes the input voltage to appear at the primary winding of the transformer as pulses at the switching frequency of between 20 and 200 kHz. The transformer output is then rectified and smoothed to produce the required stabilized DC voltage.

The inverter frequency, which is outside the audible range, is usually fixed, while the duty cycle is variable to provide the voltage level required.

Virtually all computer equipment is today powered by switch mode power supplies.

Applications

Switch mode power supplies are used in a variety of applications ranging from computers, servers, and associated equipment. They are also employed in domestic electronic equipment, security equipment, and most of the battery-operated equipment where high efficiency and small sizes are a necessity.

Advantages

There are many advantages to using switch mode technology:

- SMPS designs are more compact and use smaller transformers.
- Smaller size – an essential requirement for most electronic devices with limited space.
- Flexible technology.
- High power density.
- High efficiency of between 68% and 90%.
- Transformer-isolated supply produces a stable output that is independent of the input supply voltage.

Disadvantages

These include:

- Generation of electrical noise.
- Can be costly due to extra components.
- Complex design.
- Extra external components that also require more space.

Switch mode power supplies are considerably smaller and lighter than linear power supplies.

The images opposite show the size difference between comparable switch mode (right) and linear (left) 30 A power supplies.

Homemade Power Supplies

Of course, having learned about transformers, rectifiers, and voltage regulators, you could always design and construct your own fixed or variable supply. This is not too difficult, even for a beginner, and so makes a very useful project for anyone wishing to get more into electronics.

Hot tip

Look up designs available on the internet or in electronics magazines if you want to make your own bench supply. It is interesting and you will learn a lot.

Here are just a few examples of power supplies that have been made by home constructors. They are a good way of becoming familiar with electronics, and circuit design and construction.

All components are readily available and relatively cheap, plus you can decide if you want to use an analog meter, digital display, or simply use a multimeter for the voltage readout. Once constructed and working, you can actually use the supply to power other projects.

For the first supply (shown above), the constructor has even produced layouts so that a neat PCB can be made as described in the Chapter 10.

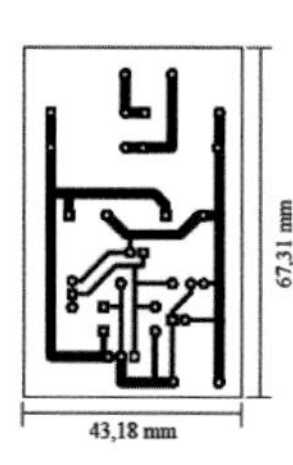

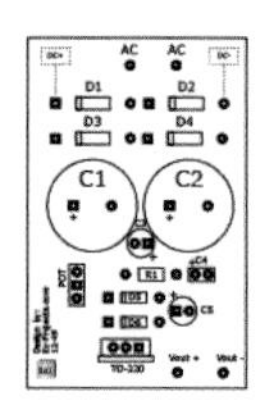

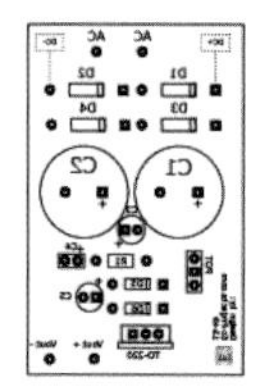

Here are other examples. Some have been made from salvaged PC power supply parts to reduce cost.

Beware

Remember that there will be high voltages present on the input side of any power supply that plugs into the mains supply.

12 Test Equipment

This chapter looks at various test instruments and their different types. Typical instruments are shown, together with many specifications. Soldering irons are also covered, as are analog and digital probes.

The Analog Multimeter

The most frequent measurements made in electronics are those of voltage, current, and resistance so it makes sense to have a single instrument that can do all of these instead of needing three separate ones.

Even the simplest of multimeters can be an extremely useful item of test equipment.

In Chapter 5 you learned that this instrument is called a *multimeter*. It can measure both AC and DC values. The name has come about because it is both a multifunction device and a multi-range meter. Some multimeters will not only measure voltage, current, and resistance but may also incorporate facilities for checking transistors, diodes, and possibly capacitors or even frequency, though this would depend on the type of meter.

Types of multimeter

There are two types of basic multimeter available. One is called *analog* because it uses an analog meter to display values, whilst the other is a *digital* multimeter employing, as the name suggests, a digital display.

Analog multimeter

An analog multimeter is the original type of multimeter and has therefore been around for many years. These multimeters are extremely flexible instruments, and invaluable for finding faults in electronic circuits.

Start by first selecting a high range then work downwards if you are unsure of the level of the signal you are measuring.

Meter sensitivity

The analog meter used could cause issues when it comes to the sensitivity of the instrument. This is because the meter needs to draw a small amount of current from the circuit it is measuring in order for the meter to deflect. As such, the meter appears as another resistor placed between the points being measured. This resistance is usually referenced in terms of kΩ per volt, and from this the effective resistance can be calculated for any given range.

For example, if a multimeter had a sensitivity of 20 kΩ per volt, then on the range having a full scale deflection of 30 volts, it would appear as a resistance of 30 x 20 kΩ; i.e. 600 kΩ.

To ensure the load placed on the measurement point is minimal, the resistance of the meter should be, at the very least, 10 times the resistance of the circuit being measured.

Normally, the sensitivity of an analog meter is much less on AC than DC. A meter with a DC sensitivity of 20 kΩ per volt on DC may, in fact, only have a sensitivity of 8 kΩ per volt on AC.

Don't forget to always zero the meter on an analog instrument before checking resistance.

Multimeter ranges

Analog multimeters have a variety of measurement ranges. The maximum that the range can read is described in terms of *full scale deflection* (FSD). The best accuracy is when the meter is indicating somewhere between about a quarter and all of the FSD. To help realize this, multimeters have a variety of ranges that are chosen to be reasonably close to each other or related in some way.

A typical meter may have the following full scale deflection ranges:

- DC voltage: 2.5 V, 10 V, 25 V, 50 V, 100 V, 250 V, 1000 V.
- AC voltage: 10 V, 25 V, 50 V, 100 V, 250 V, 1000 V.
- DC current: 50 μA, 1 mA, 10 mA, 100 mA, 10 A (separate).
- Resistance: 1 Ω, 10 Ω, 100 Ω, 10,000 Ω.

Points to note about this typical specification are:

- The low-voltage AC voltage range may have a different scale to the others due to non-linear rectification issues.
- AC current is often not measured because rectification circuitry would have to be included within the multimeter.
- The high-voltage ranges, especially AC, will often use a different input lead connection so the reading can be taken through a different shunt. Also, to keep high voltages away from the rotary switch for safety reasons.
- A battery located inside the multimeter provides current for resistance measurements. No power is needed for voltage and current ranges so for these, the meter appears passive.
- The three highest resistance ranges simply multiply the 1 Ω meter reading by 10, 100, or 10,000 dependent upon the range setting. This allows for quite low resistance measurements to be made as well as very high ones. Typically, the higher resistance ranges may use more battery power than the low resistance ranges. An "OFF" position is often provided on the rotary switch to extend battery life.

Turn your meter off when you have finished using it to conserve the life of the battery.

...cont'd

Always check the battery every so often to ensure it hasn't leaked due to aging. They last a long time and get forgotten about till it's too late.

Check the battery!

Before you try making any resistance measurements it is always advisable to check the battery, especially if the meter is new, as it may not have a battery installed. Batteries are often left out of instruments to avoid them deteriorating and leaking if the device is stored for some time.

You should also periodically check your multimeter battery (or the batteries in any other test instrument you may have), as they do tend to last a long time and, if forgotten about, could end up leaking and messing up the battery compartment, contacts, or even the multimeter itself.

Measurements

Using an analog multimeter is quite easy. You may remember from Chapter 2 that voltage appears across a component, whilst current flows through a component. With this in mind, making voltage, current, and resistance measurements is fairly straightforward once you have become familiar with how to use the multimeter.

Here are some simple steps to follow when making measurements:

1. Insert the probes into the correct positions on the multimeter as there may be a number of different connections that can be used

2. Set the rotary switch to the correct range and measurement type for the measurement to be made. For safety, and especially if you are unsure of the level of what you are about to measure, it is best to select a range greater than the reading you anticipate, as it can always be reduced later if necessary. This will prevent any possible damage to the meter due to overloading

3. Carefully place the two probes across a component for a voltage reading, or temporarily disconnect one side of the component and attach the probes in series to read current

4. Note the reading, adjusting the range for a maximum deflection to get the best reading. This is the way to get the most accurate value. Carefully remove the probes

To measure current you will need to break a connection at a convenient point so that the meter can be placed in series with the item under test.

The Digital Multimeter

The digital multimeter is rapidly replacing the analog multimeter, though many analog types are still available. Some experienced electronic engineers actually prefer the analog type because the meter reading on a digital meter takes a few moments to settle.

Although the pointer on an analog meter also needs time to rise, with practice it is possible to get a quicker reading, plus you can see if the reading is fluctuating between two values, which can be difficult to achieve with a digital meter. This is when you then need an oscilloscope.

When reading the meter on an analog multimeter, looking vertically down on it will give the most accurate result.

Compare the digital meter shown opposite with the old AVO Mk 8 analog multimeter to see how technology has advanced.

However, the AVO is a precision instrument and, although getting on in years, can still be found in some workshops and is regarded as the standard for multimeters. It has a reputation that is hard to beat, and can still be found on the second-hand market. It is big but accurate and built to last!

Being small, cheap, and easy to read, the digital multimeter (or DMM) is today one of the most widely used pieces of test equipment. The DMM can provide a high degree of measurement accuracy when working on an electronics or electrical circuit.

Versatile

Like its analog counterpart, a basic DMM can measure amps, volts, and ohms, and is used in the same way. With the added possibility of being able to easily incorporate further functionality into an integrated circuit, many digital multimeters are now able to make a number of other additional measurements as well.

Use the continuity buzzer to quickly check for continuity without having to actually look at the multimeter.

Many DMMs now include functions such as frequency counter, continuity tester (incorporating a buzzer to easily check for a short or open circuit), capacitance, and temperature measurements, etc.

The DMM display is easy to see and read. Most have four digits, the first of which can often only be either a 0 or 1. There will be a +/- indication, as well as a few other smaller indicators such as AC/DC/Ω, and so on, depending upon the model of DMM.

Oscilloscope

An oscilloscope may look a complicated instrument at first, but with a little practice you will actually find it quite easy to use.

The oscilloscope was mentioned earlier. It enables you to actually see a signal as a trace on a screen, and thus makes it much easier to see any problems occurring in an electronic circuit.

The name *oscilloscope* comes from the fact that it enables oscillations to be viewed. Initially, oscilloscopes used cathode-ray tubes to display the waveforms, but now LCDs or plasma displays are used as they are smaller, more convenient to use and, in particular, do not require the very high voltages of the old CRTs.

Types of oscilloscope

Oscilloscopes are typically analog or digital. The first types were analog but now, with advances in digital technology, most are processor controlled and incorporate digital signal processing to provide improved performance and clear waveform display.

As well as the traditional stand-alone units, there are now oscilloscope devices designed to interface with computers that will do the processing before displaying the waveform on their screens. Connection is usually via USB links, but other types of interface are also available.

Building an item of test equipment is both a good way of learning how a circuit works, as well as being a good way of improving your construction skills.

For those who like to construct electronic circuits, very cheap digital oscilloscope kits that you build yourself have been available for purchase on the internet for some time. These use a small 2.3" TFT screen and produce surprisingly good results for a very low cost. All of the parts needed are supplied, including the screen and test lead, plus full construction details.

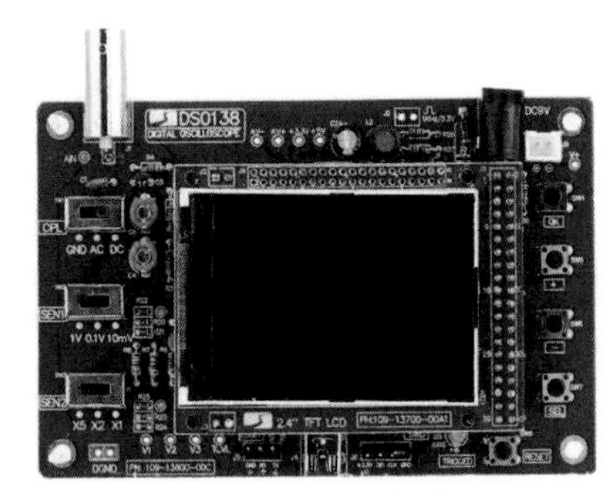

You simply need to be able to solder and carefully follow the instructions to produce a useful piece of test equipment. One such typical kit is shown here.

Specification

Shown below is a typical 50 MHz digital oscilloscope. It has four channels, internal memory for storing data, and is software driven.

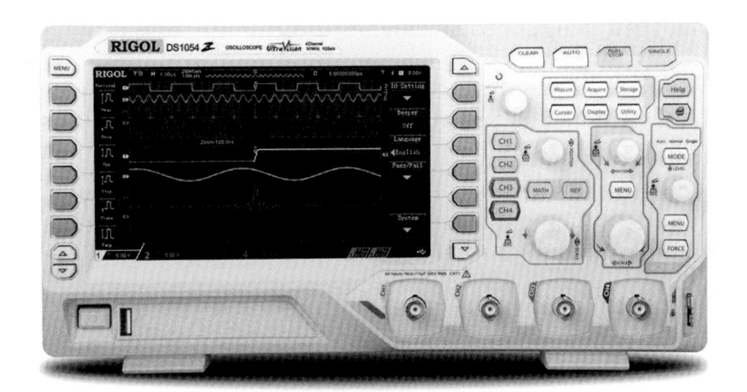

The specification for oscilloscopes can be quite varied and extensive; this can lead to possible confusion. Therefore, a basic understanding of the terms used and what they mean is very useful. In understanding the basic oscilloscope specifications, you will get to know the limitations of any given scope. This will help when it comes to actually using the oscilloscope in practice.

Even the older CRT oscilloscopes can give good results if properly set up and adjusted.

The specifications are slightly different between analog and digital oscilloscopes, although the basics like timebase range (seconds per division), upper frequencies, accuracy, etc., are essentially the same. Specific to digital oscilloscopes are items like the number of DAC bits and memory depth.

Specifications include the following main items:

- **Input range** – typical scopes offer selectable full scale input ranges from ±50 mV to ±50 V.
- **Number of channels** – two channels is normal so you can compare two signals; some scopes have four channels; whilst eight channels are available on some PC-based scopes.
- **Bandwidth** – the maximum frequency of signal that can pass through the front-end amplifiers.
- **Triggering** – to stabilize/capture repetitive waveforms.
- **Resolution** – high resolution for more accurate measurements.

Function Generator

A *function generator* is an item of electronic test equipment that is used to generate different types of electrical waveforms over a wide range of frequencies. Some of the most common waveforms produced by a function generator are the sine wave, square wave, triangular wave, or sawtooth wave.

Although function generators can cover both audio and some RF frequencies, they are not strictly suitable for applications that require very stable frequency signals. They are, however, ideally suited for the repair or testing of electronic equipment.

Beware

Not all function generators have digital displays so it may be necessary to use a frequency counter to help set the frequency.

Operation

From the simple function generator shown below, you can see that it is quite easy to operate – just follow these steps:

1. Switch on the power to the function generator
2. Select the type of wave required – sine, square, or triangular
3. Select an appropriate frequency range here
4. Adjust the frequency control till correct on the display

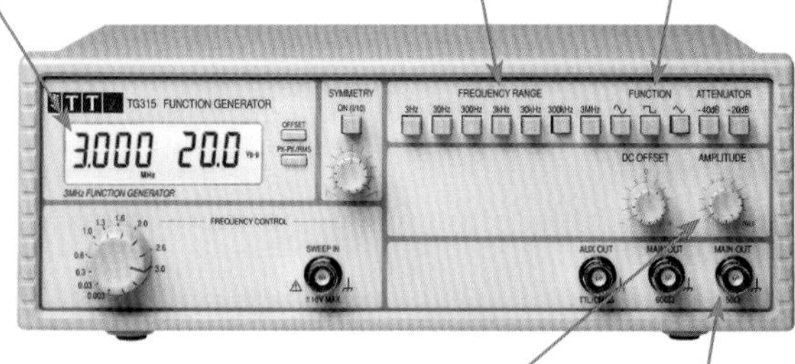

5. Set the amplitude to the required level (as per the display)
6. Connect the output lead to the MAIN OUT socket
7. Connect the other end of the output lead to the appropriate point of the circuit under test, where you want to inject the signal

Be careful not to over-drive the circuit under test. Use the attenuator if necessary.

Probes

There are different types of probes – some are just for making a connection between a test instrument and a point on a circuit, whilst others may be a complete handheld test instrument.

Simple test leads

A typical multimeter probe consists of a pair of single flexible wires: one red and one black. One end of each test lead is fitted with a connector that fits the multimeter, and on the other end is a rigid plastic section that comprises both a handle and the exposed probe contact point.

Beware

Regularly check all probe leads for breaks or exposed wires that could give you an electric shock if touched with bare skin. Immediately replace all worn probes.

Voltage probe

Used to measure voltages to a high degree of accuracy. The probe must not affect the voltage being measured, and so will present a high impedance to the circuit under test. Popular for high-voltage measurements.

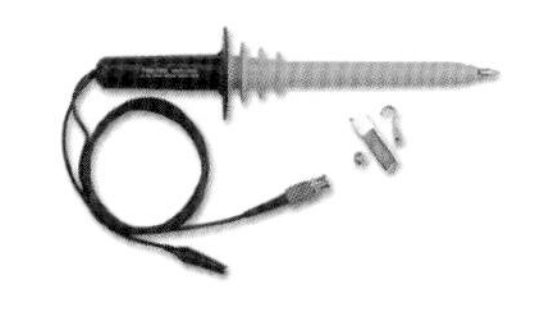

Oscilloscope probe

These fall into two main categories, passive and active; the most common being passive, meaning that they contain no active electronic parts and require no external power. Because of the signals that often need to be observed on an oscilloscope, such as high frequencies, coaxial cable is used to avoid picking up any interference and for increased accuracy. Oscilloscope probes are characterized by their frequency limit.

Some oscilloscope probes have a x10 switch so check this hasn't been accidentally moved to the wrong position if the display appears wrong.

Logic probe

Logic probes are used for testing of digital circuits. They are cheap, easy to use, and can provide a simple way of testing slow-moving digital logic levels and signals. The red and green indicator LEDs are used to display:

- A logic HIGH state.
- A logic LOW state.
- Digital pulses, as indicated by the brightness of the LEDs.

Frequency Counter

A *frequency counter* is a device that measures a periodic signal and determines its frequency in hertz. The result is displayed on a digital readout; the number of digits used depending on the maximum frequency the counter can go up to.

Like a lot of test equipment, the frequency counter is very easy to use, especially with recent advances in technology that have allowed features such as high accuracy and auto-ranging to be incorporated into today's devices. Frequency counters are widely used in a variety of areas to measure the frequency of repetitive signals, from audio frequencies right up to the microwave range.

Accuracy

Some frequency counters, like the one below, can have a very wide frequency range available within one single instrument. Because the accuracy of the frequency count is affected by the sensitivity of the input circuitry, it is not really feasible to cover the complete range (0.01 Hz to 2.4 GHz in this case) by using just one input.

The accuracy of a frequency counter readout is dependent on the purity of the signal being measured.

Most instruments, therefore, will have at least two signal inputs: the first, input A, for the frequency range 0.01 Hz-50 MHz, and input B for the range 50 MHz-2.4 GHz. In this way, the two input circuits can be designed for maximum accuracy in their individual frequency ranges.

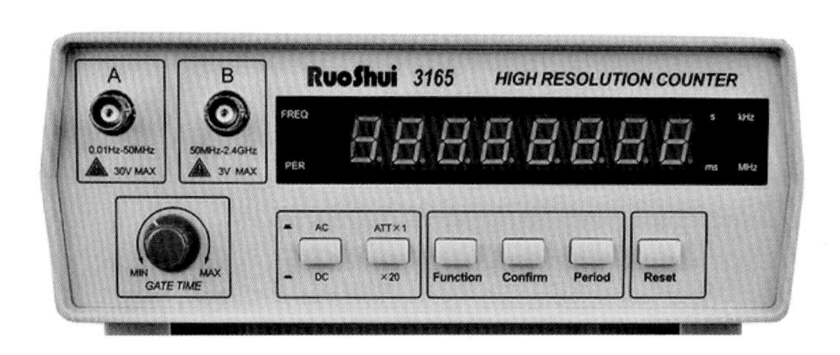

Using a frequency counter

Very good fast results and accurate measurements are normally obtained from frequency counters. It is normally just necessary to plug the instrument in, turn it on, and connect it to the points where the frequency measurement need to be made.

Most frequency counters can now also be used to measure digital logic signals. To avoid putting a load on the circuit under test, counters normally have a high impedance input – typically, 1 MΩ.

RF Signal Generator

As the name suggests, *radio frequency* or *RF signal generators* are used for testing or developing high-frequency circuits. By applying a signal with known characteristics such as frequency, amplitude, and modulation from the signal generator to the circuit, it's possible to examine the response of the circuit and see exactly how well it is working.

Like the function generator, make sure you don't over-drive the circuit under test.

The RF signal generator is very similar to the function generator, but designed specifically for the various requirements of radio frequency circuits and the accuracy needed for microwaves.

Basic modules

A modern RF signal generator consists of a number of main sections:

- **Oscillator** – typically a very stable frequency synthesizer. Can be set to the required frequency by commands from the controller.
- **Amplifier** – a special circuit to amplify the output from the oscillator module to a set level; needs to maintain the output level accurately at all frequencies and temperatures.
- **Control** – uses advanced processors to control all aspects of the operation of the instrument.
- **Attenuator** – used on the output of the signal generator to maintain source impedance and signal level accuracy.

Functions

A large variety of functions and facilities are offered by modern RF signal generators:

- **Frequency range** – must be able to cover a wide range of high frequencies including any harmonics.
- **Output level** – controlled to a relatively high degree of accuracy. Range is normally limited at the top end by the final amplifier in the RF signal generator.
- **Modulation** – applies modulation to the output signal. Signal generators that support complex modulation are often referred to as vector signal generators.
- **Control** – through traditional front panel controls or possibly via remote control options such as RS-232, Ethernet, and USB (for many of the latest instruments).

The modulation function is normally used when testing communications equipment that uses frequency modulation.

The Soldering Iron

Although a *soldering iron* isn't strictly an item of test equipment, it is indeed a very important item. It is basically a hand tool used to supply heat to melt solder so that it can flow and joint the leads of electronic components together or to a circuit board.

It is composed of an insulated handle and a heated metal tip. The heating is caused by passing an electric current through a resistive heating element situated just before the metal tip. Some soldering irons are temperature controlled to ensure the correct amount of heat is used and components aren't damaged through overheating whilst being soldered.

Different types

There are a number of different types of soldering irons available:

- **Simple soldering iron** – a low-power iron with a rating of 15 W to 35 W. Used for general electrical and electronics work. Primarily suitable only for making soldered connections between component leads or small circuit board track. Not suitable for things with large thermal capacity such as connections to a metal chassis or heatsink.

  Simple irons normally run at an uncontrolled temperature, so their ability to melt solder may be reduced as the temperature drops when heating something large.

- **Temperature-controlled soldering iron** – to overcome temperature drops, many irons used in electronics incorporate a temperature sensor and some method of temperature control to keep the tip temperature constant.

  Various means are used to control temperature. The simplest uses a variable power control, whilst another type of system uses a thermostat inside the iron's tip to automatically switch power on and off to the element, so maintaining the temperature. Another approach is to use a magnetized soldering tip that loses its magnetic property at a specific temperature. Whilst

When soldering, only use the minimum amount of heat and solder to make a nice, clean joint. There are many soldering tutorials and soldering hints and tips online, should you need further information on soldering.

Always use the correct size of tip to ensure sufficient heat gets to the joint. Poor joints eventually become dry joints and cause faults.

Temperature-controlled irons usually allow you to use different-sized tips or replace the tip when it has worn out.

the tip is magnetic, it closes a switch to supply power to the heating element. When the tip exceeds the design temperature, the magnetism is lost and the switch contacts open. Power is cut off and the tip begins to cool until the temperature drops enough to restore magnetization sufficiently to close the contacts and restore power again. Power will be in the region of 100 W; tips are often replaceable.

- **Soldering station** – consists of an electrical power supply, a facility for user adjustment of temperature, and a temperature-controlled soldering iron or soldering head. The hot iron sits in a stand when not in use, and can be cleaned on a wet sponge.

  It can be used for a number of functions such as soldering electronic components or a rework station. For work on surface-mount components there will also be a hot air gun, a vacuum pickup tool, and a soldering head. For use as a desoldering station there will be a desoldering head with a vacuum pump for desoldering any through-hole components.

Soldering stations are expensive but make removing and replacing faulty components easy, and minimize any possible damage to the circuit board.

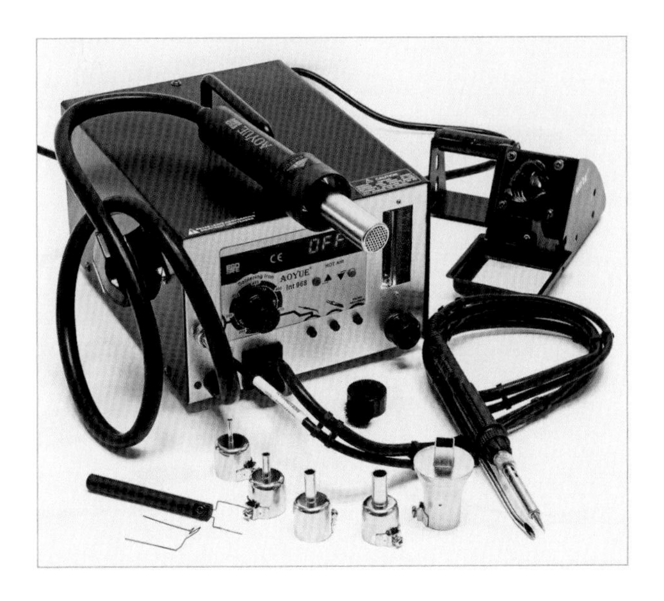

- **Soldering tweezers** – for soldering and desoldering small surface-mount components with two terminals, such as resistors, capacitors, and diodes. They can be either free-standing or controlled from a soldering station, and have two heated tips mounted on arms that you squeeze gently together.

Soldering tweezers are invaluable for working with SMT components.

Logic analyzers can be tricky to set up due to the large number of probes that may need connecting to a circuit.

Logic Analyzer

A *logic analyzer* is an electronic instrument used to capture and display multiple signals from a digital circuit or system. A logic analyzer may convert the captured data into a timing diagram that allows the user to see if the signals are correctly synchronized in time, and if pulses are clocking correctly, etc.

Logic or digital circuits operate at very high speeds. Normally, oscilloscopes can't keep up with such signals and so have difficulty accurately displaying high-speed pulses. Also, you often need to view a number of different traces in one time frame; something that an oscilloscope has difficulty with.

Logic analyzers are designed with advanced triggering capabilities and are useful when a user needs to see and compare the timing relationships between the many high-speed signals in a digital system. They can display multiple signal inputs simultaneously.

Different types of logic analyzers

A number of different analyzers are available, from stand-alone to small USB capture devices that feed the data to a computer for display on the screen. Typical examples are shown here.

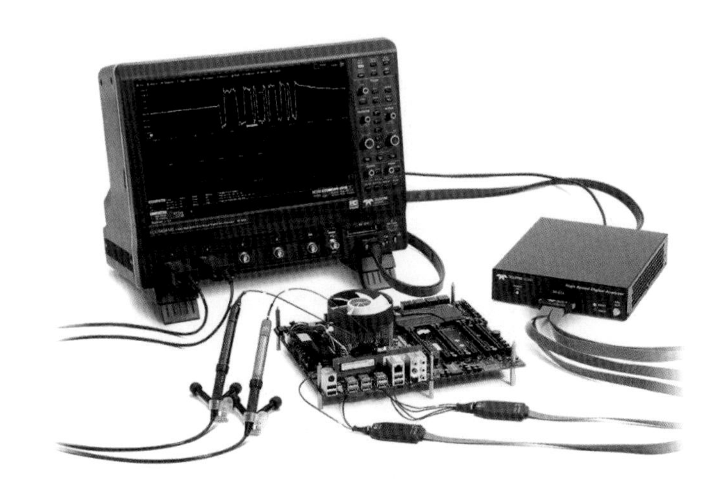

USB signal capture devices that use computers to do the signal processing and displaying make a cheap logic analyzer.

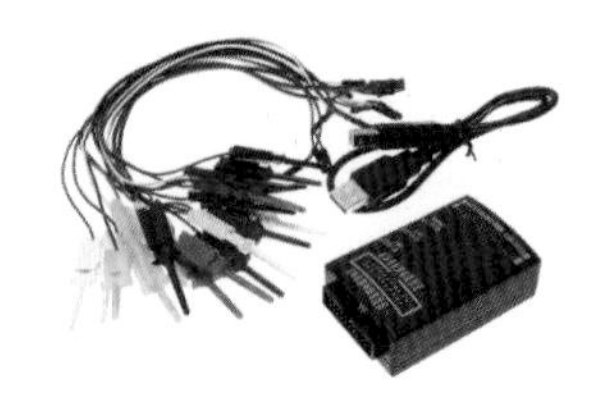

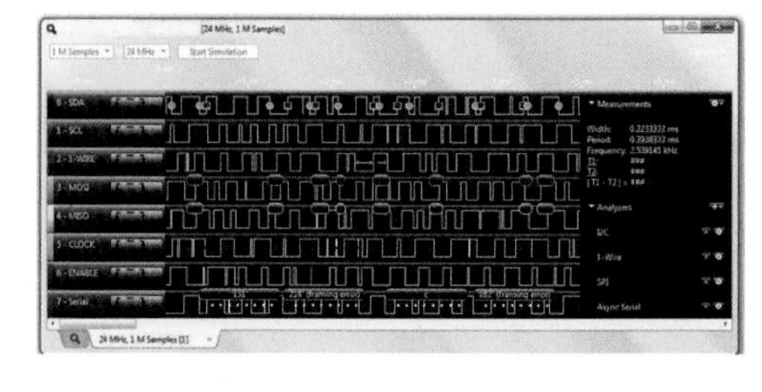

13 Digital Electronics

Here, you will learn all about digital circuits and logic gates. You are introduced to the basic logic gates that are the building blocks for digital circuits, and will learn how a truth table is used to identify the function of a gate or logic circuit.

Digital Logic

Also known as *Boolean logic*, *digital logic* is the term used to define the concept of making extremely complicated decisions based on simple "yes/no" questions. This is the fundamental concept underpinning all modern digital circuits, software controlled devices, and computer systems.

Simply put, it is the system of rules that allow us to design and control complex circuits. The physical part of digital logic is called the *hardware*; the programming part is called the *software*. Not all digital circuits require software to function, but they do all work on a true or false principle that can be represented by something called a *truth table* – more of which later.

Don't forget

Digital logic is a completely different aspect of electronics than that you have been learning so far. Signals have only two states: "On" or "Off".

Digital logic is also the underlying system that drives electronic circuit board design. It also refers to the manipulation of binary values (0 or 1) using *logic gates* to construct circuits that can perform relatively simple tasks or highly complex multiple instructions at the speed of light, such as implemented in the operation of a computer system.

Digital circuits

Today we are surrounded by digital logic circuits, or *digital circuits* as they are known. They can be found in most appliances around the house – from the simple toaster through to the large flat-screen television. Like a computer, the larger devices are software controlled. The TV, for example, is now a mostly digital device. It receives digital signals; a built-in processor and software converts these into pictures and sends them for display on the screen.

Because it is easier to process digital signals than analog ones, communications equipment such as the radio amateur transceiver shown here is really a computer with a radio front end. Received analog signals are converted to digital and cleaned up for better reception.

Hot tip

Digital circuits are versatile and can be used in many applications where some form of control is required.

Logic Gates

A circuit that has been designed to perform a basic logic function is called a *logic gate*. The most common types of logic gates are listed in the table below, together with their standard symbols.

The ANSI/MIL symbols are traditional ones, but the IEC and British Standard symbols are commonly used by design packages worldwide, as they are easier to draw using software. Logic gate building blocks are used extensively in computer and other digital circuits.

Gate	ANSI/MIL	IEC	British BS 3939
Buffer		1	1
NOT		1	1
AND		&	&
NAND		&	&
OR		≥1	≥1
NOR		≥1	≥1
XOR		=1	=1
XNOR		=1	=1

ANSI/MIL – American National Standards Institute/Military standard.
IEC – International Electrotechnical Commission.

It is not good practice to mix logic gate symbols from different standards when drawing circuits.

Logic gates are the building blocks for digital logic circuits.

...cont'd

Logic is easy to understand: a logic state can only ever be 0 or 1.

Logic states

The signals used in logic gates can only be one of two states: either On or Off. These states are often referenced in different ways, such as Yes or No and True or False, but only ever in two opposite states. The On state is represented by a 1 and the Off state by a 0, hence the terms *logic 0* or *logic 1*.

This, therefore, means that inputs and output of any logic gate can only ever represent one of two states – a 1 (known as a high state) or a 0 (also known as a low state).

Binary

A number expressed using a two-state numbering system is called a *binary number*. It consists purely of the digits 0 and 1, and each digit in the number is called a binary *bit*. A group of eight binary bits is called a *byte*. Larger groups of bytes are normally binary multiples (16, 32, 64, etc.) and are often referred to as a *word*.

Buffers

All the gates shown in the table on page 149 provide logic functions except for the buffer. The job of a buffer is to give the input signal a bit of a boost. Sometimes, the output of a gate may have to drive the inputs on a large number of other gates, and so the current from the output could do with strengthening – this is where the buffer comes in. It always gives the same state on the output as that on its input. For example, a 1 on the input results in a 1 on the output, and vice versa.

Logical functions

A logical function is exactly what it says: the output of a gate will give the logical result of the states of its inputs, depending on what basic logic function that gate has been designed to do.

Logic functions are just that: logical results from a defined logic input.

For example, the NOT gate is so called because the output will always NOT be the input. The logic of this is that with only two possible states, if a 1 is on the input of the gate then it will not be a 1 at output, and not a 1 is a 0! Similarly, a 0 at the input will not be a 0 at the output, so it has to be a 1 – simple as that.

The way other logic gates work follows similar logical thinking – the names tell you the logic used; i.e., AND, NAND (Not AND), OR, NOR (Not OR), etc. Some people think of the NOT gate as an inverter as it swaps the signal around, but this is not good practice.

AND Gate

Truth tables

Remembering that the two binary levels are often thought of as representing true or false, it follows that an easy way of showing the function of a logic gate or circuit is to use a truth table. You simply draw up a table listing all the possible input combinations and the resulting output state. The same method is used when you have a circuit consisting of a number of logic gates connected together; you just need a bigger table to also show the output of each gate so you can work out the eventual output result.

Truth tables show exactly how a logic gate or circuit functions.

The AND gate

Let us now see how the AND gate works. It is so called because it will only produce a logic 1 at its output when all the inputs are at a logic 1. To help you with this, just think in terms of needing a logic 1 on the first input AND a logic 1 on the second input to get a logic 1 at the output.

AND gate – output is 1 only when BOTH inputs are 1.

The AND gate has at least two inputs but only one output. Some AND gates have three, four, or eight inputs but still only one output. No matter how many inputs, the rule for seeing a 1 at the output is still the same – there needs to be a 1 on the first input and a 1 on the second input and a 1 on the third input, etc., etc. – hence the name AND gate. Here is the two-input truth table:

Input A	Input B	Output	AND Gate
0	0	0	A, B
0	1	0	
1	0	0	
1	1	1	

Always make sure you identify the pin layout when working with logic chips, as they all look very similar.

So many of the basic gates are used that they are available in integrated circuit packages containing more than one gate. This is a quad two-input AND gate chip. When building a circuit, any unused gates are simply ignored.

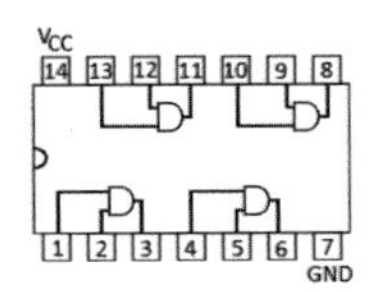

NAND Gate

Sometimes you may want the opposite of the output of a gate. Adding a NOT function to the gate output would achieve this but it means an extra gate; this may not be feasible if space on a circuit board is tight. Instead of a logic 0 you now get a logic 1 output, and instead of a logic 1 you now get a logic 0 output.

Never connect two logic gate outputs together.

To make things simpler, a logic gate with a NOT function already fitted to the output was designed. To identify this, the letter N was added to the gate name; i.e., NOT AND becomes NAND.

The NAND Gate

This gate performs a NOT AND function in a single circuit. The truth table for the NAND gate is shown here.

Input A	Input B	Output	NAND Gate
0	0	1	
0	1	1	A, B
1	0	1	
1	1	0	

You will notice that although the input pattern remains the same, the output is the opposite of that of the AND gate. In other words, when both inputs are a 1, the output is now a 0. For the other three input combinations, where an AND gate would give a 0 at the output, the NAND gate gives a 1. Logically speaking, this gate is truly not an AND gate!

NAND gate – output is 1 if BOTH inputs are 0.

The 7400 quad NAND gate in a military-grade metal flat package (7400W) was the first product in the series brought out by Texas Instruments in October 1964. This was followed a couple of years later by the extremely popular commercial-grade plastic DIL device (7400N).

Universal gate

The NAND gate is also known as a *universal gate* because it can be configured to do the function of any of the other logic gates. For example, with both inputs strapped together you actually get a NOT gate function. Check this out on the truth table!

OR Gate

You don't always need a gate that tells you when all of the inputs have gone to a 1. Sometimes you need the output to become a 1 when just one of the inputs switches from a 0 to a 1; this is what the OR gate was designed to do.

The OR gate

In the same way that the AND gate got its name from the logical way it functions, the same is true for the OR gate. It is so called because it will give a logic 1 at the output when you have a logic 1 on the first input OR a logic 1 on the second input OR a logic 1 on both inputs – hence the name OR gate. When both inputs are at a 0, the output is also a 0. Below is the OR gate truth table.

OR gate – output is 1 if AT LEAST one input is 1.

Input A	Input B	Output	OR Gate
0	0	0	
0	1	1	A, B
1	0	1	
1	1	1	

You will have noticed by now that all of the symbols covered so far follow a set convention: inputs are on the left and outputs are to the right. It is good practice to stick to this convention when drawing logic diagrams.

While inputs can be connected together, outputs should never be connected to one another, only to other inputs. However, one output may also be connected to multiple inputs, though sometimes a buffer will be required if it is too heavily loaded, as already mentioned.

An OR gate has at least two inputs, though gates with three, four, or more inputs are also available.

Things to note about the ANSI symbols:

- The function is not always printed, as the symbol is assumed to be sufficient identification.
- The A-B input notation is standard but isn't shown on circuits.
- Two-input devices are standard, though some devices have more.

NOR Gate

As with the AND gate, there are times when you require the opposite output to that appearing on the output of an OR gate, so the NOT OR or NOR gate was developed.

The NOR gate

This gate performs a NOT OR function in a single circuit. The truth table for the NOR gate is shown here.

Input A	Input B	Output	NOR Gate
0	0	1	
0	1	0	A B
1	0	0	
1	1	0	

Again, the input pattern remains the same as for the OR gate, but the output is the opposite. In other words, when both inputs are a 0, the output is now a 1. For the other three input combinations, where an OR gate would give a 1 at the output, the NOR gate gives a 0. Logically speaking, this gate is not an OR gate!

With combinational logic, state changes are instant, but with sequential logic, states might only change when activated by a clock signal.

Combinational logic

Along with the AND, OR, NAND, etc., the NOR gate belongs to the combinational logic category – the other category being sequential logic.

Combinational logic changes instantly, which means that the output of the circuit responds as soon as the input changes. You need to allow for some delay whilst signals propagate through the circuit elements, but this takes hardly any time at all as it is in the order of nanoseconds.

A clock signal is a square wave used to coordinate the switching of logic states in digital circuits. Clocking occurs on an edge, the point at which the clock waveform changes from a high to a low state (falling edge) or from a low to a high state (rising edge).

Sequential logic

Sequential circuits generally have a clock signal. The logic state changes propagate through the various stages of the circuit on edges of the clock. Typically, a sequential circuit will be built up of blocks of combinational logic linked by "memory elements" to hold some of the logic states until activated by a clock signal.

XOR Gate

The OR gate provides a useful function but it does have one disadvantage. Although the output goes to a 1 when either of the inputs goes from a 0 to a 1, it also gives a 1 out when both inputs are at a 1. This is not always convenient.

XOR gate – output is 1 if ONLY one input is 1.

The XOR gate

This gate is named from the fact that its output is a 1 only when one of the inputs goes to a 1. If both inputs are at a 1 then the output stays at a 0. In other words, the output is a 1 when only one of the inputs is exclusively a 1, hence eXclusive OR. You can see this exclusive function from the truth table.

Input A	Input B	Output	XOR Gate
0	0	0	
0	1	1	A
1	0	1	B
1	1	0	

The basic XOR gate only has two inputs. There are exclusive OR gates with more than two inputs, but how they will function depends on their implementation. In most cases, an XOR gate will output a 1 if an odd number of its inputs are a 1. However, you can see that this behavior doesn't follow the strict definition of XOR, which states that only one input must be exclusively a 1.

The logic symbol ⊕ is used to denote XOR in algebraic expressions.

You can build any logic circuit or function using only NAND gates.

If no XOR chips are to hand then an XOR gate circuit can be made from four NAND gates – the so called *universal gate*. The circuit for this is shown below. Why not work out the truth table for this, and prove that it does function as an XOR?

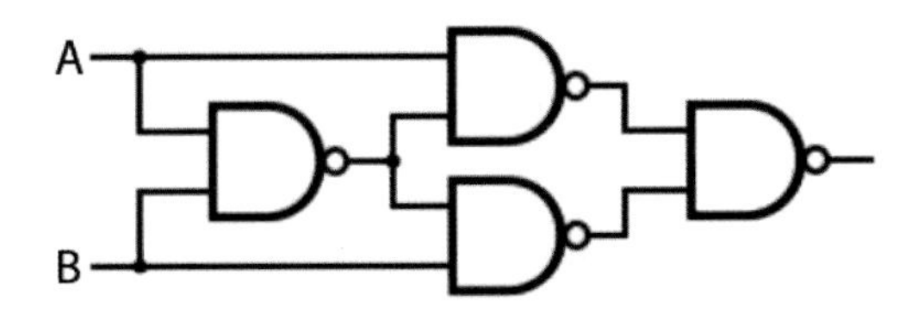

XNOR Gate

The XNOR gate was designed for use when the opposite output was required from an XOR gate. This saved adding an extra NOT gate to the output of an XOR, and hence increasing the gate count.

The XNOR gate

This gate performs a NOT XOR function in a single circuit. The truth table for the XNOR gate is shown here.

XNOR gate – output is a 0 if ONLY one input is a 0.

Input A	Input B	Output	XNOR Gate
0	0	1	A, B (XNOR gate symbol)
0	1	0	
1	0	0	
1	1	1	

The input pattern remains the same as for the XOR gate, but the output is the opposite. When both inputs are a 0 or a 1, the output is now a 1. For the other two input combinations, where a XOR gate would give a 1 at the output, the XNOR gate gives a 0. Logically speaking, this gate is not an XOR gate!

The XNOR function can also be constructed using five NOR gates. Can you work out how this is possible?

The XNOR function can be constructed using just five NOR gates.

Summary

To recap, combinational circuits are built of seven basic logic gates:

- NOT gate – output is always the opposite of the input.
- AND gate – output is 1 if BOTH inputs are 1.
- NAND gate – output is 1 if BOTH inputs are 0.
- OR gate – output is 1 if AT LEAST one input is 1.
- NOR gate – output is 1 if AT LEAST one input is 0.
- XOR gate – output is 1 if ONLY one input is 1.
- XNOR gate – output is a 0 if ONLY one input is a 0.

Boolean Expressions

The inventor of Boolean algebra was George Boole, the son of a shoemaker and born in 1815 in Lincoln, England. Obviously, this was long before electronics as we know it and logic circuits were created. His interest was mathematics, and he did it to show that true or false could be expressed as a formula.

Boolean algebra is ideal for expressing a logic function as a formula.

Boolean algebra and the truth table mentioned earlier are simply a mathematical description of logic and logic functions. Using Boolean algebra it is possible to write out very complex logic expressions, but you must take care with the NOT symbols or the expression could end up being more complex than is necessary.

Remember that if you NOT NOT a 0 you will end up with a 0 again; same for a 1. You are just putting two NOT gates in series.

You can write down all of the logical functions in this chapter mathematically, using Boolean algebra, to create a Boolean expression. There are only three symbols involved so this is not as complex as it sounds. Some examples are given in the table below.

Symbol	Logic Function	Example	NAND Gate	Logic Gate
•	AND	$A \cdot B$	A AND B	2-input AND
		$A \cdot B \cdot C$	A AND B AND C	3-input AND
+	OR	$A + B$	A OR B	2-input OR
—	NOT	$\overline{A}$	NOT A	The output of a NOT
		$\overline{A \cdot B}$	A NAND B	2-input NAND
		$\overline{A + B + C}$	NOT (A OR B OR C)	3-input NOR

De Morgan's laws

Being based on logic, large and complex Boolean expressions can often be reduced to less complex expressions by manipulating the logic! The British mathematician and logician Augustus De Morgan formulated very simple laws for this very purpose. First, change all AND to OR, and OR to AND, NOT the individual AND and OR terms then NOT the whole expression. Because a double NOT cancels, with a little manipulation the example $\overline{A+\overline{B}+\overline{C}}$ reduces to $\overline{A}\cdot B\cdot C$ – give it a try and see for yourself!

Use De Morgan's laws to reduce complex Boolean expressions to much simpler ones.

Logic Chips

Years ago, logic gates were made using discrete components – that is, individual resistors, capacitors, transistors, etc. With advances in integrated circuit and microchip technology, manufacturers eventually started producing IC chips with however many of one gate they could squeeze into the package. For example, six (or Hex) NOT gates fit inside a standard 14-pin IC package.

Don't forget

Dual in-line (DIL) packages are available in 14-pin or 16-pin format.

Typically, standard logic gates are available in 14-pin or 16-pin DIL (dual in-line) chips. As just mentioned, the number of gates per IC depends on the number of inputs per gate and how many pins there are in the package. Two input gates are common, though three, four, and eight are also available. The greatest number of inputs on a single gate is on the 74133 13-input NAND gate, which uses a 16-pin package. The following is just a small selection of the many ICs in production.

7400-series

The most well known group of IC logic chips is the original 7400-series made by Texas Instruments, with the prefix "SN"; i.e., SN7400. The series became very popular, and other manufacturers quickly released compatible devices with the same pin configuration. The 7400 sequence was retained to aid identification, though many manufacturers used different prefixes or no prefix at all.

7400 Quad 2 input NAND Gates

7402 Quad 2 input NOR Gates

7408 Quad 2 input AND Gates

7432 Quad 2 input OR Gates

7486 Quad 2 input XOR Gates

747266 Quad 2 input XNOR Gates

74133 Single 13 input NAND Gate

7404 Hex NOT Gates

The small notch on the top surface of a logic chip indicates where pin 1 starts.

Technology improved, and the 74xx family expanded. Low power (74L), high speed (74H), high-speed Schottky (74S), and low-power Schottky (74LS) are just a few of the additions. The 7400-series uses transistor-transistor logic (TTL), though CMOS logic gates soon followed (4000-series).

The 7400-series chips were the first IC logic chips to be introduced.

14 Circuits & Reference

Now it is time to put theory into practice. This chapter details a few circuits for you to build – from a simple starter exercise to projects you can use in the electronics workshop. Handy reference tables are provided for designing your own circuits.

LED Tester

It is now time to start putting what you have read about so far into practice. This chapter lists a selection of circuits that are both useful and fun to build.

These circuits have been compiled from different sources and so contain a number of regional variations for the circuit symbols. This will give you good practice in becoming familiar with the preferred way of drawing electronic circuit diagrams in different countries. You should, however, be easily able to read the diagrams.

Hot tip

Starting with a simple circuit like this helps you get used to circuit construction, handling components, and seeing how they work.

It is always best to start with something simple, so why not see how an LED works with this easy-to-make LED tester. You don't need a fancy printed circuit board or even stripboard. For this first attempt at constructing a circuit, breadboard will do or, if you do not have one to hand, you can always just twist the wires together whilst making sure nothing shorts out!

Circuit diagram

This is the circuit diagram. The anode and cathode identification for the LED has been included to get you started, and so you don't connect the LED the wrong way round.

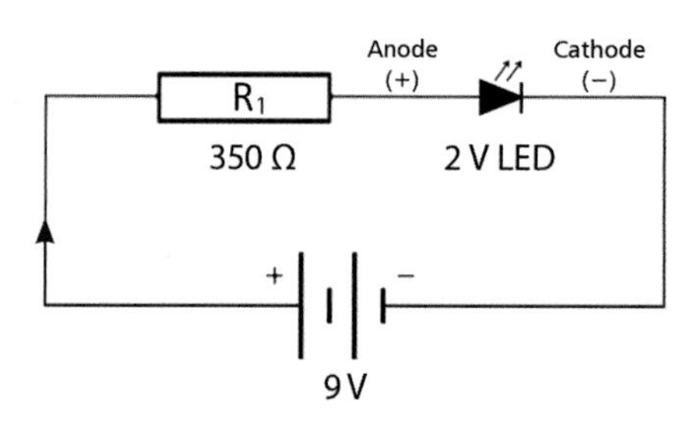

Always check the polarity of a component and ensure you connect it the correct way round.

The anode is the long pin and is connected to the biasing resistor R_1 as it is to the positive side of the 9 V battery. The shorter pin of the LED (the cathode) is connected to the negative battery terminal. You must connect the LED correctly for it to work. You can experiment and try different LEDs and values of R_1 to see the result, but don't allow the LED to draw too much current and fail.

Parts list

- R_1 – 350 Ω resistor.
- LED – 2 V type, any color.
- Battery – 9 V (or a power supply set to 9 V).

Crystal Set

The crystal set is the earliest form of radio receiver. It is easy to make and needs no power supply to work, as it derives its energy from the received signal. However, it does need a good, long piece of wire for an aerial and a sensitive earpiece. To drive a loudspeaker you would need to add an audio amplifier stage like the one described on page 164.

Good results can be had with the circuit shown here. Remember that it was the crystal set that introduced the public in general to radio broadcasting. They became extremely popular and many thousands were sold. Lots of hobbyists built their own following the basic circuit shown. You even wind the coil yourself.

Simple as it is, a crystal set can give surprisingly good results, even if you don't live very close to a radio transmitter.

Circuit diagram

The circuit is easy to follow and only uses five components. Details on how to wind the coil are given in the notes below.

Parts list

- Coil or inductor
- Variable capacitor
- OA91 diode
- A fixed capacitor
- Crystal earpiece

Aerial
Crystal Earpiece
Germanium Diode
Coil wound on Ferrite Rod
Variable Capacitor
Capacitor
Earth

Build notes

You can substitute any other germanium-type diode in place of the OA91. Note that a silicon diode will not work – it must be a germanium diode.

Modern variable capacitors tend to be around 150 pF maximum in value and are readily available.

To match the variable capacitor value given, make the coil by winding 75 turns of 30 SWG enameled copper wire on a ferrite rod of about 8 mm diameter and 100 mm length. If you don't have 30 SWG wire, use something similar; even fine PVC insulated wire can be used. You may need to add or remove turns to get the desired coverage for your homemade crystal set.

A small fixed capacitor of about 330 pF is used to remove the radio signal and leave just the audio signal for the earpiece.

Use the longest aerial you can and a good connection to ground/ earth for best results.

555 Timer Circuit

The 555 timer is a popular and commonly used small integrated circuit, designed to produce a variety of accurately timed output waveforms with the addition of a few external components and a resistor/capacitor network.

Chip designer Hans Camenzind invented the 555 timer but it almost didn't get made because his employer believed it was easy to make timers using cheap discrete components, so why spend money on manufacturing a chip to do the same job!

Simplified block diagram

To help understanding of how the 555 timer works, a simplified block diagram of the internal circuitry is shown here. The timer gets its name from the fact that it uses three internally connected 5 kΩ resistors to generate the two comparators' reference voltages.

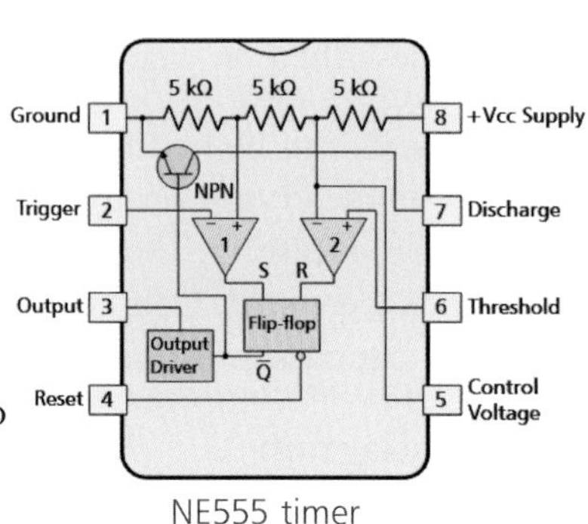

NE555 timer

Despite its cheapness, the 555 timer IC is a useful precision timing device that can act as a simple timer to generate single pulses or long time delays.

NE 555 pin configuration

Two different packages of the 555 timer are shown; both work in the same way. The 8-pin DIL device will be used for an LED flasher circuit to show how the timer works.

- Pin 1 – Ground supply.
- Pin 2 - Triggering input, start of timing input.
- Pin 3 – Timer output signal.
- Pin 4 – Active low reset input.
- Pin 5 – Control voltage, controls comparator thresholds.
- Pin 6 –Threshold end of timing input.
- Pin 7 – Open collector output to discharge timing capacitor.
- Pin 8 – NE555 chip supply voltage is 4.5 V to 16 V (VCC).

The SE555 IC can take a maximum supply of 18 V. It will give TTL compatible output and up to 200 mA current.

The NE555 timer chip was first introduced in 1971 by Signetics and was an immediate hit. It is still used today, and billions have been manufactured and sold.

A simple LED flasher circuit can be made using the NE555 timer.

Circuit diagram

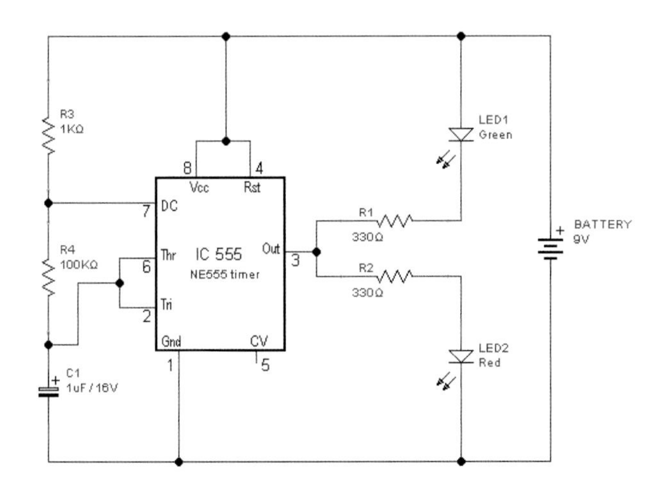

In this circuit, the timer is functioning as an astable multivibrator, meaning that it is free-running.

Circuit description

Two LEDs are connected to the timer output. You will notice from the way they are connected in the circuit that the anode of LED1 connects directly to the positive supply line, and the cathode of LED2 connects directly to the negative line.

This means that when the output of the 555 timer pulses positive, it will drive LED2 to turn On whilst LED1 stays Off. When the output of the 555 timer pulses negative, it will drive LED1 to turn On whilst LED2 turns Off. Because pin 4 (Reset) is connected to Vcc, no reset occurs and the timer oscillates freely, causing LED1 and LED2 to continue flashing alternately.

The frequency of oscillation is controlled by the timing components R3, R4, and C1. Try changing their values and see how it affects the speed of flashing.

Why not use a variable resistor for R3 or R4 and see what happens as you increase and decrease the resistance value.

Parts list

- Resistors – R1 & R2 (330 Ω), R3 (1 kΩ), R4 (100 kΩ).
- Capacitor – C1 1μF.
- LED – 2 V red 2 V green.
- Sundries – NE555 timer, 9 V battery.

Audio Amplifiers

Before integrated circuits came along you had to build an audio amplifier using valves or transistors and a number of other discrete components. If you needed any real power then the transistors would be quite large and probably mounted on heatsinks.

All of that changed when the audio amplifier chip was introduced. Now it was possible to have a complete amplifier circuit capable of easily driving a small loudspeaker all in one chip.

The LM386 was primarily designed for a 4 Ω speaker load, but is rated for 8 Ω and 32 Ω loads as well.

There are many types of audio amplifier ICs, such as the LM series from Texas Instruments.

Part number	Description
LM380	2.5W audio power amplifier (fixed 34dB gain)
LM384	5W audio power amplifier (fixed 34dB gain)
LM386	Low voltage audio power amplifier
LM833	Dual high speed audio amplifiers

LM386

This device has been around since 1983 and can still be found in low-power, battery-driven applications all around the world. It is perfect for building your first chip-based audio amplifier, as it runs from a single power supply, is efficient, and needs no heatsink. The configuration below has a gain of 20.

Circuit diagram

Only eight components are needed to make the amplifier, and that includes the IC and loudspeaker. The output is 0.3 W, and is perfectly adequate for most purposes. The LM386N is very cheap so this project can be built for hardly any cost.

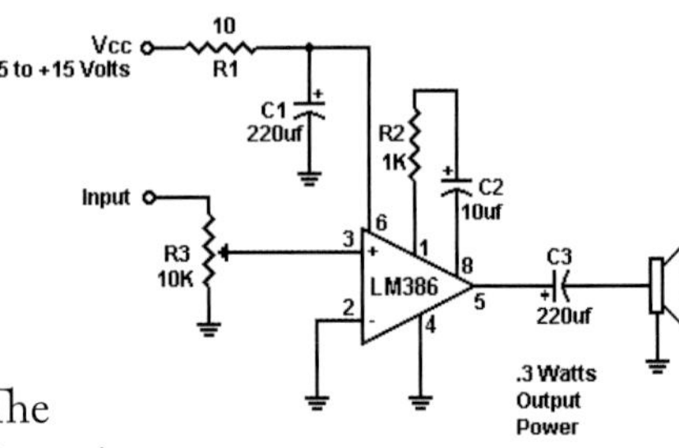

Output capacitor coupling is mandatory in just about all IC audio amplifier designs (Capacitor C3 here).

Parts list

- Resistors – R1 (10 Ω), R2 (1 kΩ), R3 (100 kΩ).
- Capacitors – C1 (220 μF), C2 (10 μF), C3 (220 μF).
- Integrated circuit – LM386 (LM386N-1 = 300 milliwatts);
  (LM386N-3 = 700 milliwatts);
  (LM386N-4 = 1000 milliwatts).
- Sundries – Loudspeaker (4 Ω or 8 Ω), 9 V battery;
  Breadboard or stripboard, connecting wire.

The LM386 is a low-power device. Should a higher-power output be required then the TDA series can be used.

Some of these chips were produced in their millions and are still available for incredibly little cost despite becoming obsolete.

TDA series audio amplifiers

The TDA integrated circuit series is popular for amplifier designs and projects that require significantly higher power than the LM series can deliver. TDA audio amplifier circuits are produced by a number of manufacturer, the most widely used ICs being the TDA2030 and TDA2003 for small audio amplifier kits, and TDA7294 for higher power amplifiers.

TDA2003

This 5-pin audio amplifier IC has been around for a while now and is capable of providing up to 10 W into a 2 Ω load and 6 W into a 4 Ω load when powered at 14.4 V. It is quite a robust device due to in-built short circuit protection, and can withstand a permanent short circuit on the output as long as the supply voltage doesn't exceed 16 V.

Whilst the maximum operating DC voltage is 18 V, the TDA2003 can withstand a voltage of up to 28 V maximum without sustaining damage. It also features an integrated thermal limiting circuit. Originally designed for car audio – hence the single supply of about 12 V – the TDA2003 proves to be a good option for small power amplifiers. Although it is considered obsolete now, there are plenty of electronic parts suppliers that still have the TDA2003 in stock and at very low prices.

The TDA2003 makes an extremely cheap and useful audio amplifier.

Circuit diagram

The basic circuit diagram for the 10 W amplifier is shown here. Component values are not critical and are those recommended for the original car radio audio amplifier design. It is advisable to use 25 V electrolytic capacitors for C1 and C2, and small non-polarized film-type instead of ceramic for the others.

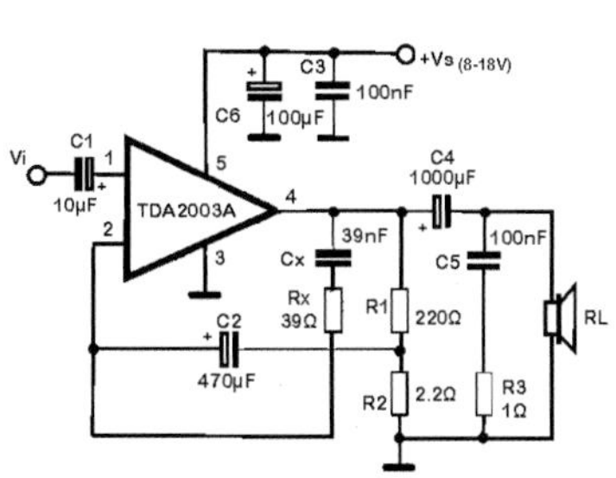

To keep heat to a minimum when heavily driven, the TDA2003 should be mounted on a heatsink. Why not build two of these to make a 10 W per channel amplifier for an MP3 player?

...cont'd

Hot tip

If you want even more power output, why not look at the datasheet for the TDA2030A?

Datasheet

Datasheets have already been mentioned in Chapter 9. As well as containing full specification for a component or device, they also usually have suggested applications circuits and typical component values. The sample from the TDA2003 datasheet shown here gives you some idea of their usefulness.

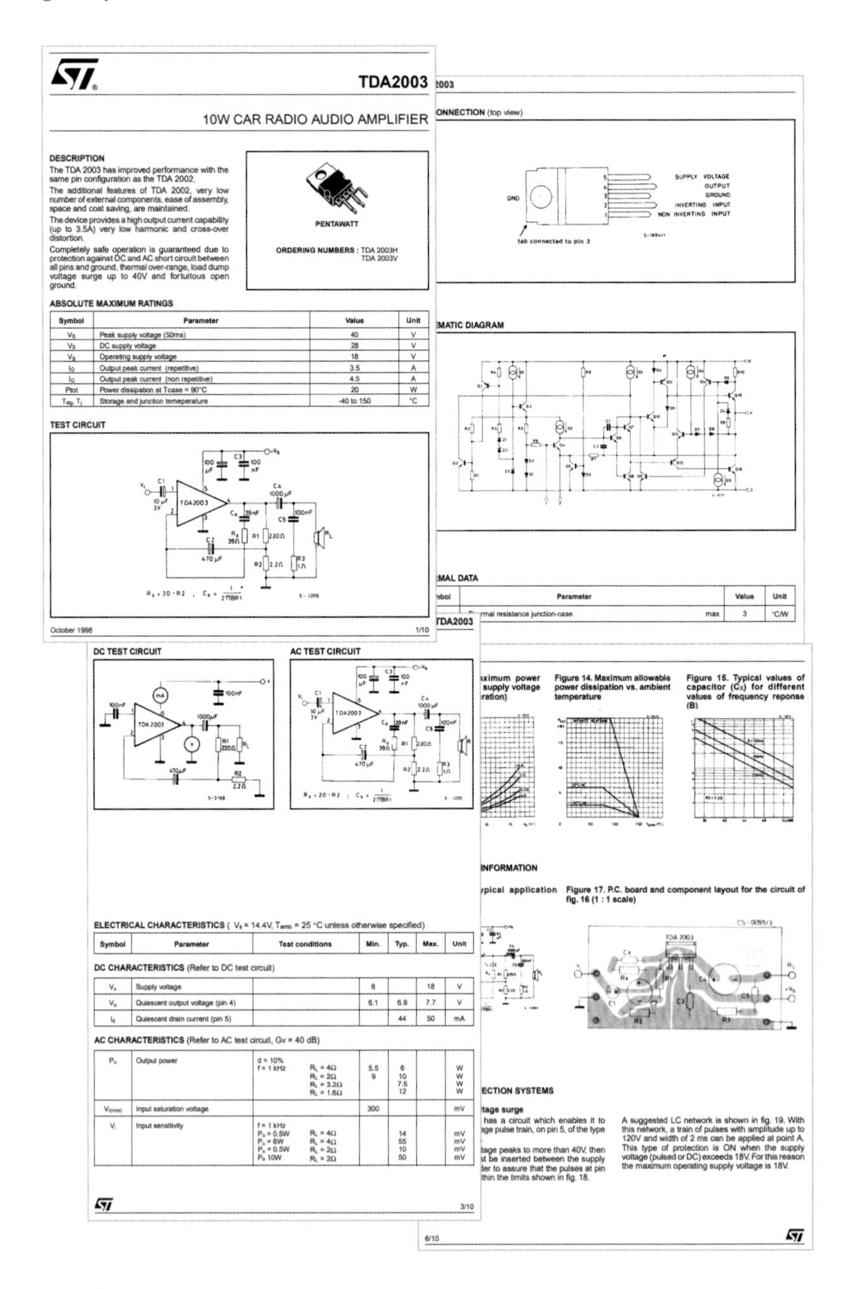

TDA2003

10W CAR RADIO AUDIO AMPLIFIER

DESCRIPTION

The TDA 2003 has improved performance with the same pin configuration as the TDA 2002.

The additional features of TDA 2002, very low number of external components, ease of assembly, space and cost saving, are maintained.

The device provides a high output current capability (up to 3.5A) very low harmonic and cross-over distortion.

Completely safe operation is guaranteed due to protection against DC and AC short circuit between all pins and ground, thermal over-range, load dump voltage surge up to 40V and fortuitous open ground.

PENTAWATT

ORDERING NUMBERS : TDA 2003H
TDA 2003V

ABSOLUTE MAXIMUM RATINGS

Symbol	Parameter	Value	Unit
V_S	Peak supply voltage (50ms)	40	V
V_S	DC supply voltage	28	V
V_S	Operating supply voltage	18	V
I_O	Output peak current (repetitive)	3.5	A
I_O	Output peak current (non repetitive)	4.5	A
Ptot	Power dissipation at Tcase = 90°C	20	W
T_{stg}, T_j	Storage and junction temeperature	-40 to 150	°C

TEST CIRCUIT

October 1998 1/10

2003

ONNECTION (top view)

SUPPLY VOLTAGE
OUTPUT
GROUND
INVERTING INPUT
NON INVERTING INPUT

tab connected to pin 3

EMATIC DIAGRAM

MAL DATA

bol	Parameter		Value	Unit
	rmal resistance junction-case	max	3	°C/W

TDA2003

DC TEST CIRCUIT

AC TEST CIRCUIT

ELECTRICAL CHARACTERISTICS (V_s = 14.4V, T_{amb} = 25 °C unless otherwise specified)

Symbol	Parameter	Test conditions	Min.	Typ.	Max.	Unit
DC CHARACTERISTICS (Refer to DC test circuit)						
V_s	Supply voltage		8		18	V
V_o	Quiescent output voltage (pin 4)		6.1	6.9	7.7	V
I_d	Quiescent drain current (pin 5)			44	50	mA
AC CHARACTERISTICS (Refer to AC test circuit, Gv = 40 dB)						
P_o	Output power	d = 10% f = 1 kHz R_L = 4Ω R_L = 2Ω R_L = 3.2Ω R_L = 1.6Ω	5.5 9	6 10 7.5 12		W W W W
$V_{i(rms)}$	Input saturation voltage		300			mV
V_i	Input sensitivity	f = 1 kHz P_o = 0.5W R_L = 4Ω P_o = 6W R_L = 4Ω P_o = 0.5W R_L = 2Ω P_o 10W R_L = 2Ω		14 55 10 50		mV mV mV mV

3/10

ximum power supply voltage ration)

Figure 14. Maximum allowable power dissipation vs. ambient temperature

Figure 15. Typical values of capacitor (C_X) for different values of frequency reponse (B)

INFORMATION

ypical application

Figure 17. P.C. board and component layout for the circuit of fig. 16 (1 : 1 scale)

ECTION SYSTEMS

tage surge

has a circuit which enables it to age pulse train, on pin 5, of the type

tage peaks to more than 40V, then st be inserted between the supply der to assure that the pulses at pin thin the limits shown in fig. 18.

A suggested LC network is shown in fig. 19. With this network, a train of pulses with amplitude up to 120V and width of 2 ms can be applied at point A. This type of protection is ON when the supply voltage (pulsed or DC) exceeds 18V. For this reason the maximum operating supply voltage is 18V.

6/10

Datasheets are readily available online, even for obsolete devices. Simply search by device number and view the PDF file.

Power Supply

A power supply is one of the most useful items of equipment you can have. Instead of buying one, the Multisim example on page 116 shows the circuit for a variable power supply that you could build instead. It is based on the LM317 voltage regulator that has also been covered on pages 104 and 116.

However, it is now also possible to buy a complete power supply kit from a number of sources on the internet so cheaply that it is hardly worth spending time on designing a circuit board and sourcing components.

A typical kit, also based on the LM317, is shown here. As you can see, it is complete with all of the components, PCB, and hardware included. It even has digital readout of voltage and current. The fully regulated voltage output is variable between 1.25 V and 12 V. The see-through case even lets you admire your handiwork! All you have to do is carefully solder everything together; there is no setting up to do as the digital readout module is pre-configured.

The kit

Here are various pictures of the kit. If you would like to tackle this project, just search "variable power supply kit" on the internet and have fun building something that you can use for years.

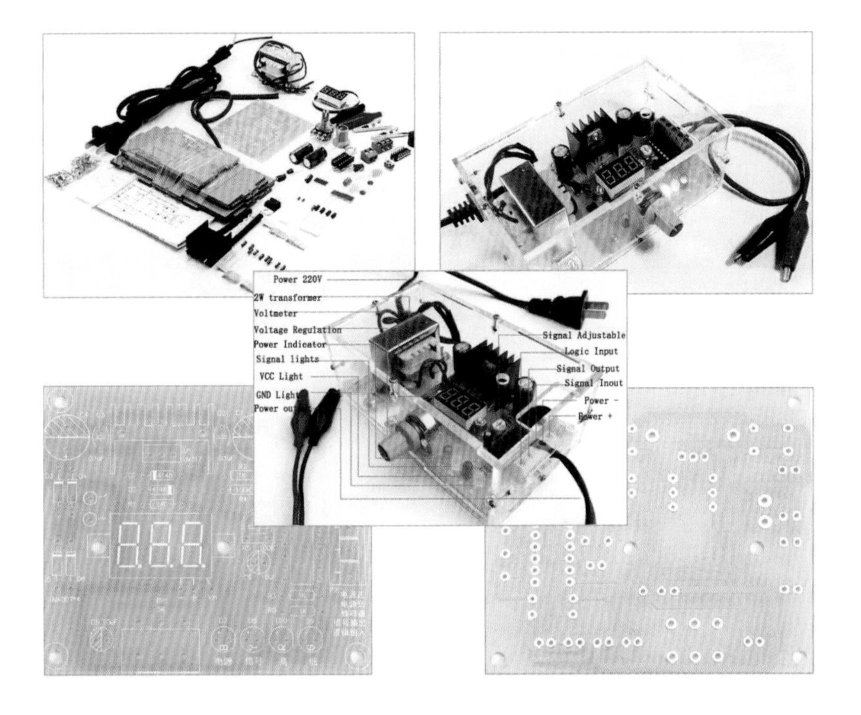

For your own safety do take care when building and using any equipment that will connect to the mains supply.

You can also buy a digital multimeter kit with a built-in transistor tester for about the same price.

Why not add a USB downconverter that you can then use for charging portable devices.

Common Transistors

Use this reference as a useful aid to choosing suitable transistors when designing a circuit. It lists the specifications for transistors commonly used in small electronic projects and circuits.

The list also gives the application that a particular transistor was designed for, the type (NPN or PNP), the maximum current capability (I_C), the power dissipation (P_D), and the maximum frequency the transistor is suitable for (F_T).

Hot tip

Hundreds of different transistors have been manufactured over the years, but many have similar specifications. Use these tables to help you choose a suitable alternative transistor if the one you need isn't currently available.

BC Series source - SourcForge.com

Type	Package	NPN/PNP	Ic (mA)	Pd (mW)	Vce (max)	Vcb (max)	hfe	@ Ic	FT (MHz)	Complement	Replacement	Application
BC107	TO-18	NPN	100	300	45	50	100-450	2	300	BC177		General
BC108	TO-18	NPN	100	300	20	30	110-800	2	300			General
BC109	TO-18	NPN	100	300	20	30	200-800	2	300			General
BC109C	TO-18	NPN	100	300	20	30	420-800	2	300			General
BC142	TO-39	NPN	1.0 A	800	60	80	20-60	200	80	BC143		High Pwr Audio
BC143	TO-39	PNP	1.0 A	800	60	60	20-40	300	-	BC142		High Pwr Audio
BC177	TO-18	PNP	100	300	45	50	75-260	2	150	BC107		High Pwr Audio
BC178	TO-18	PNP	100	300	25	30	75-500	2	150			High Pwr Audio
BC179	TO-18	PNP	100	300	20	25	125-500	2	150			High Pwr Audio
BC182L	TO-92	NPN	100	350	50	60	120-500	2	150	BC212L		High Pwr Audio
BC183L	TO-92	NPN	100	350	30	45	120-800	2	150	BC213L		High Pwr Audio
BC184L	TO-92	NPN	100	350	30	45	250-800	2	150	BC214L		High Pwr Audio
BC212L	TO-92	PNP	100	350	50	60	60	2	280	BC182L		High Pwr Audio
BC213L	TO-92	PNP	100	350	30	45	80-400	2	350	BC183L		High Pwr Audio
BC214L	TO-92	PNP	100	350	30	45	140-600	2	320	BC184L		High Pwr Audio
BC307	TO-92	PNP	100	350	45	50	160-460	2	280			Amplifier
BC327	TO-92	PNP	500	625	45	50	100-600	100	100	BC337		Audio
BC328	TO-92	PNP	500	625	25	30	100-600	100	100			Audio
BC337	TO-92	NPN	500	625	45	50	100-600	100	100			Audio
BC338	TO-92	NPN	500	625	25	30	100-600	100	100			Audio
BC546	TO-92	NPN	100	500	65	80	110-450	2	300			High Pwr Audio
BC547	TO-92	NPN	100	500	45	50	110-800	2	300			High Pwr Audio
BC548	TO-92	NPN	100	500	30	30	110-800	2	300			High Pwr Audio
BC549	TO-92	NPN	100	500	30	30	200-800	2	300			Low Noise
BC556	TO-92	NPN	100	500	80	80	75-475	2	200			High Pwr Audio
BC557	TO-92	PNP	100	500	45	50	75-800	2	200			High Pwr Audio
BC558	TO-92	PNP	100	500	30	30	75-800	2	200			High Pwr Audio
BC559	TO-92	PNP	100	500	30	30	125-800	2	200			High Pwr Audio
BC639	TO-92	NPN	1.0 A	1.0 W	80	100	40-250	150	130			Audio
BC640	TO-92	PNP	1.0 A	1.0 W	80	100	40-250	150	50			Audio

The BC series transistors are usually found in circuits originating from Europe and further east.

2N Series source - SourcForge.com

Type	Package	NPN/PNP	Ic (mA)	Pd (mW)	Vce (max)	Vcb (max)	hfe	@ Ic	FT (MHz)	Complement	Replacement	Application
2N2222	TO-92	NPN	800	500	40	75	100-300	150	150	2N2907		Switching
2N2222	TO-18	NPN	800	500	40	75	100-300	150	300	2N2907		Switching
2N2905	TO-39	PNP	600	600	40	60	100-300	150	200			Switching
2N2907	TO-39	PNP	600	400	40	60	100-300	150	200	2N2222		Switching
2N3703	TO-92	PNP	200	300	30	50	30-150	50	100	2N3705		High Pwr Audio
2N3704	TO-92	NPN	800	360	30	50	100-300	50	100	2N3702		High Pwr Audio
2N3904	TO-92	NPN	200	350	40	60	100-300	10	300	2N3906		Switching
2N3906	TO-92	PNP	200	350	40	40	100-300	10	250	2N3904		Switching
2N4401	TO-92	NPN	600	350	40	60	100-300	150	250	2N4403		High Pwr Audio
2N4403	TO-92	PNP	600	350	40	40	100-300	150	200	2N4401		High Pwr Audio
2N5401	TO-92	PNP	600	625	150	160	60-240	10	100			Amplifier

The SN series transistors are usually favored for circuits originating from the USA and Canada.

C Series source - SourcForge.com

Type	Package	NPN/PNP	Ic (mA)	Pd (mW)	Vce (max)	Vcb (max)	hfe	@ Ic	FT (MHz)	Complement	Replacement	Application
C1815	TO-92	NPN	400	600	70	70	120	-	200			High Pwr Audio
C2233	TO-220	NPN	4.0 A	75 W	300	300	25	-	4			High Current
C2330	TO-16	NPN	100	900	300	300	100	-	50			HV Video
C2482	TO-16	NPN	100	900	300	300	100	-	50			HV Video
C9012	TO-92	PNP	500	625	20	40	120	-	200			High Pwr Audio
C9013	TO-92	NPN	500	625	20	40	120	-	200			High Pwr Audio
C9014	TO-92	NPN	500	500	40	75	200	-	300			High Pwr Audio
C9015	TO-92	PNP	500	500	40	75	200	-	300			High Pwr Audio

Prefix class

The list below explains the European transistor prefix; e.g., "BC" means general purpose, small-signal, silicon transistor.

European (EECA) source - SourcForge.com

Prefix Class	Type and Usage	Example	Equivalent	Reference
AC	Germanium small-signal transistor	AC126	NTE102A	
AD	Germanium AF power transistor	AD133	NTE179	
AF	Germanium small-signal RF transistor	AF117	NTE160	
AL	Germanium RF power transistor	ALZ10	NTE100	
AS	Germanium switching transistor	ASY28	NTE101	
AU	Germanium power switching transistor	AU103	NTE127	
BC	Silicon small-signal transistor (General use)	BC584	2N3904	
BD	Silicon power transistor	BD139	NTE375	
BF	Silicon RF (high frequency) BJT or FET	BF245	NTE133	
BS	Silicon switching transistor (BJT or MOSFET)	BS170	2N7000	
BL	Silicon high frequency high power (for transmitters)	BLW60	NTE325	
BU	Silicon high voltage (for CRT horizontal deflection circuits)	BU2520A	NTE2354	
CF	Gallium Arsenide Microwave Transistor (MESFET)	CF739	—	
CL	Gallium Arsenide Microwave power transistor (FET)	CLY10	—	

The European transistor prefix class makes it easy to see at a glance what particular usage a transistor is designed for.

Transistor Case and Pin-out

This chart supplements the transistor pin-out information detailed in Chapter 8. This is a more complete list and covers a variety of different device types, including the old metal can-type transistors that can still be found in use today, such as the popular BC109.

Beware

Always check the pin-out to ensure you don't insert a transistor the wrong way round in a circuit board.

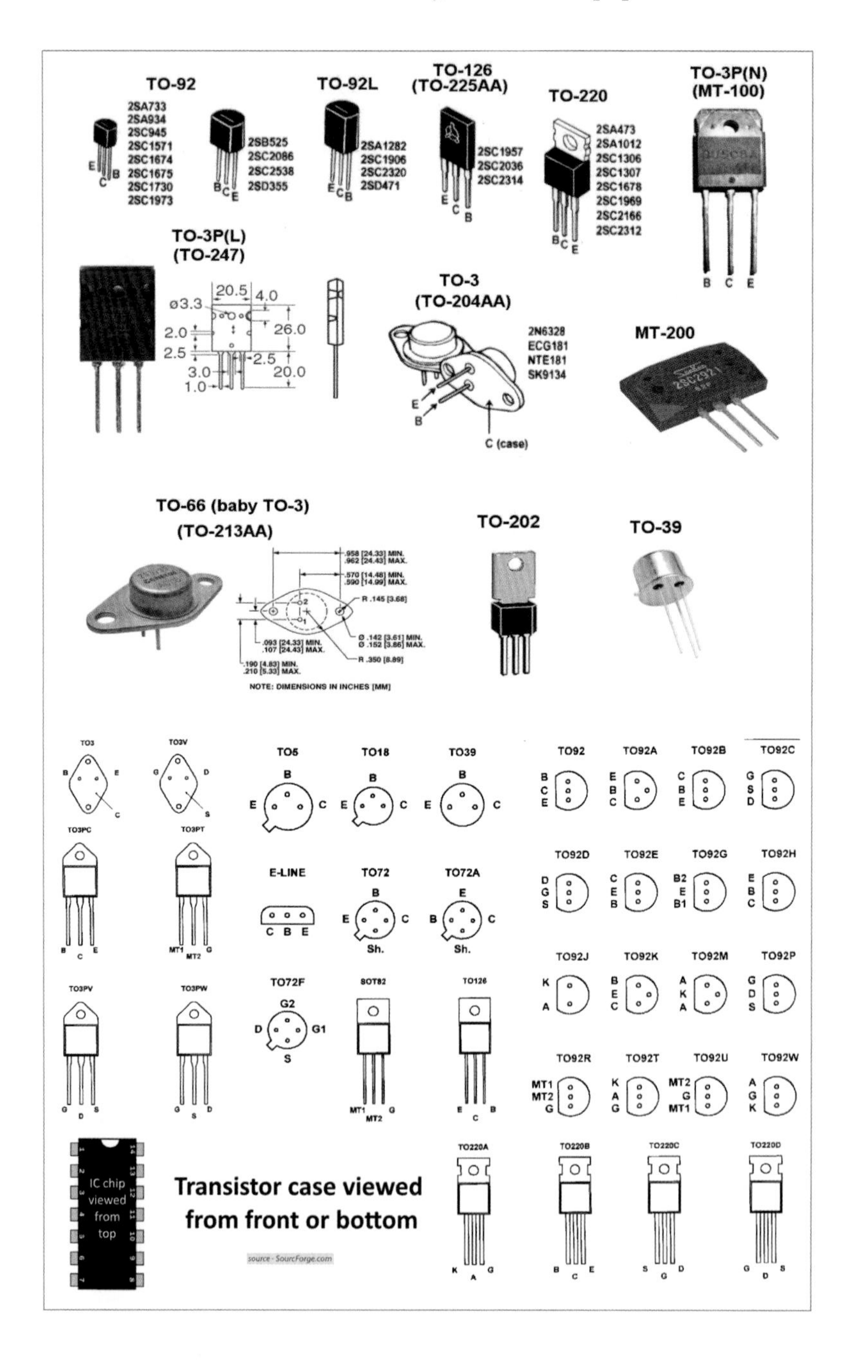

Don't forget

Transistor cases are always viewed from the bottom or the front. IC chips are always viewed from the top.

Capacitor Values

When it comes to labeling capacitors, manufacturers often have a problem as the components can often be too small to effectively stamp a value on them. Sometimes a number will fit, but there may not be enough space for the unit to be marked on there as well, such as μF, pF, or nF. If μ, p or n fits then F is implied.

Because of this, you will come across many very small capacitors marked with just a number, and even then that value will be truncated. It is quite common to see just .01 marked instead of 0.01, for example. Experience will often give you an idea as to the value indicated by these digits, usually from the type and/or size of the component, but if you are new to this, the lack of full identification can leave you very puzzled.

With low-value capacitors getting smaller and smaller due to better manufacturing processes, the markings have to be kept small too. The following table will help identify capacitors with small markings. For example, it may only say .3n on a 300 pF (0.3 nF) capacitor.

If you are having difficulty identifying a capacitor value then there are many helpful guides on the internet.

uF / MFD	pF / MMFD	nF		uF / MFD	pF / MMFD	nF
1uF / MFD	1000000pF (MMFD)	1000nF		0.001uF / MFD	1000pF (MMFD)	1nF
0.82uF / MFD	820000pF (MMFD)	820nF		0.00082uF / MFD	820pF (MMFD)	0.82nF
0.8uF / MFD	800000pF (MMFD)	800nF		0.0008uF / MFD	800pF (MMFD)	0.8nF
0.7uF / MFD	700000pF (MMFD)	700nF		0.0007uF / MFD	700pF (MMFD)	0.7nF
0.68uF / MFD	680000pF (MMFD)	680nF		0.00068uF / MFD	680pF (MMFD)	0.68nF
0.6uF / MFD	600000pF (MMFD)	600nF		0.0006uF / MFD	600pF (MMFD)	0.6nF
0.56uF / MFD	560000pF (MMFD)	560nF		0.00056uF / MFD	560pF (MMFD)	0.56nF
0.5uF / MFD	500000pF (MMFD)	500nF		0.0005uF / MFD	500pF (MMFD)	0.5nF
0.47uF / MFD	470000pF (MMFD)	470nF		0.00047uF / MFD	470pF (MMFD)	0.47nF
0.4uF / MFD	400000pF (MMFD)	400nF		0.0004uF / MFD	400pF (MMFD)	0.4nF
0.39uF / MFD	390000pF (MMFD)	390nF		0.00039uF / MFD	390pF (MMFD)	0.39nF
0.33uF / MFD	330000pF (MMFD)	330nF		0.00033uF / MFD	330pF (MMFD)	0.33nF
0.3uF / MFD	300000pF (MMFD)	300nF		0.0003uF / MFD	300pF (MMFD)	0.3nF
0.27uF / MFD	270000pF (MMFD)	270nF		0.00027uF / MFD	270pF (MMFD)	0.27nF
0.25uF / MFD	250000pF (MMFD)	250nF		0.00025uF / MFD	250pF (MMFD)	0.25nF
0.22uF / MFD	220000pF (MMFD)	220nF		0.00022uF / MFD	220pF (MMFD)	0.22nF
0.2uF / MFD	200000pF (MMFD)	200nF		0.0002uF / MFD	200pF (MMFD)	0.2nF
0.18uF / MFD	180000pF (MMFD)	180nF		0.00018uF / MFD	180pF (MMFD)	0.18nF
0.15uF / MFD	150000pF (MMFD)	150nF		0.00015uF / MFD	150pF (MMFD)	0.15nF
0.12uF / MFD	120000pF (MMFD)	120nF		0.00012uF / MFD	120pF (MMFD)	0.12nF
0.1uF / MFD	100000pF (MMFD)	100nF		0.0001uF / MFD	100pF (MMFD)	0.1nF
0.082uF / MFD	82000pF (MMFD)	82nF		0.000082uF / MFD	82pF (MMFD)	0.082nF
0.08uF / MFD	80000pF (MMFD)	80nF		0.00008uF / MFD	80pF (MMFD)	0.08nF
0.075uF / MFD	75000pF (MMFD)	75nF		0.000075uF / MFD	75pF (MMFD)	0.075nF
0.07uF / MFD	70000pF (MMFD)	70nF		0.00007uF / MFD	70pF (MMFD)	0.07nF
0.068uF / MFD	68000pF (MMFD)	68nF		0.000068uF / MFD	68pF (MMFD)	0.068nF
0.06uF / MFD	60000pF (MMFD)	60nF		0.00006uF / MFD	60pF (MMFD)	0.06nF
0.056uF / MFD	56000pF (MMFD)	56nF		0.000056uF / MFD	56pF (MMFD)	0.056nF
0.05uF / MFD	50000pF (MMFD)	50nF		0.00005uF / MFD	50pF (MMFD)	0.05nF
0.047uF / MFD	47000pF (MMFD)	47nF		0.000047uF / MFD	47pF (MMFD)	0.047nF
0.04uF / MFD	40000pF (MMFD)	40nF		0.00004uF / MFD	40pF (MMFD)	0.04nF

An LCR (inductance L, capacitance C, resistance R) meter is handy to have for identifying values you are unsure of.

...cont'd

uF / MFD	pF / MMFD	nF		uF / MFD	pF / MMFD	nF
0.039uF / MFD	39000pF (MMFD)	39nF		0.000039uF / MFD	139pF (MMFD)	0.039nF
0.033uF / MFD	33000pF (MMFD)	33nF		0.000033uF / MFD	33pF (MMFD)	0.033nF
0.03uF / MFD	30000pF (MMFD)	30nF		0.00003uF / MFD	30pF (MMFD)	0.03nF
0.027uF / MFD	27000pF (MMFD)	27nF		0.000027uF / MFD	27pF (MMFD)	0.027nF
0.025uF / MFD	25000pF (MMFD)	25nF		0.000025uF / MFD	25pF (MMFD)	0.025nF
0.022uF / MFD	22000pF (MMFD)	22nF		0.000022uF / MFD	22pF (MMFD)	0.022nF
0.02uF / MFD	20000pF (MMFD)	20nF		0.00002uF / MFD	20pF (MMFD)	0.02nF
0.018uF / MFD	18000pF (MMFD)	18nF		0.000018uF / MFD	18pF (MMFD)	0.018nF
0.015uF / MFD	15000pF (MMFD)	15nF		0.000015uF / MFD	15pF (MMFD)	0.015nF
0.012uF / MFD	12000pF (MMFD)	12nF		0.000012uF / MFD	12pF (MMFD)	0.012nF
0.01uF / MFD	10000pF (MMFD)	10nF		0.00001uF / MFD	10pF (MMFD)	0.01nF
0.009uF / MFD	9000pF (MMFD)	9nF		0.000009uF / MFD	9pF (MMFD)	0.009nF
0.0082uF / MFD	8200pF (MMFD)	8.2nF		0.0000082uF / MFD	8.2pF (MMFD)	0.0082nF
0.008uF / MFD	8000pF (MMFD)	8nF		0.000008uF / MFD	8pF (MMFD)	0.008nF
0.007uF / MFD	7000pF (MMFD)	7nF		0.000007uF / MFD	7pF (MMFD)	0.007nF
0.0068uF / MFD	6800pF (MMFD)	6.8nF		0.0000068uF / MFD	6.8pF (MMFD)	0.0068nF
0.006uF / MFD	6000pF (MMFD)	6nF		0.000006uF / MFD	6pF (MMFD)	0.006nF
0.0056uF / MFD	5600pF (MMFD)	5.6nF		0.0000056uF / MFD	5.6pF (MMFD)	0.0056nF
0.005uF / MFD	5000pF (MMFD)	5nF		0.000005uF / MFD	5pF (MMFD)	0.005nF
0.0047uF / MFD	4700pF (MMFD)	4.7nF		0.0000047uF / MFD	4.7pF (MMFD)	0.0047nF
0.004uF / MFD	4000pF (MMFD)	4nF		0.000004uF / MFD	4pF (MMFD)	0.004nF
0.0039uF / MFD	3900pF (MMFD)	3.9nF		0.0000039uF / MFD	3.9pF (MMFD)	0.0039nF
0.0033uF / MFD	3300pF (MMFD)	3.3nF		0.0000033uF / MFD	3.3pF (MMFD)	0.0033nF
0.003uF / MFD	3000pF (MMFD)	3nF		0.000003uF / MFD	3pF (MMFD)	0.003nF
0.0027uF / MFD	2700pF (MMFD)	2.7nF		0.0000027uF / MFD	2.7pF (MMFD)	0.0027nF
0.0025uF / MFD	2500pF (MMFD)	2.5nF		0.0000025uF / MFD	2.5pF (MMFD)	0.0025nF
0.0022uF / MFD	2200pF (MMFD)	2.2nF		0.0000022uF / MFD	2.2pF (MMFD)	0.0022nF
0.0022uF / MFD	2000pF (MMFD)	2nF		0.000002uF / MFD	2pF (MMFD)	0.002nF
0.0018uF / MFD	1800pF (MMFD)	1.8nF		0.0000018uF / MFD	1.8pF (MMFD)	0.0018nF
0.0015uF / MFD	1500pF (MMFD)	1.5nF		0.0000015uF / MFD	1.5pF (MMFD)	0.0015nF
0.0012uF / MFD	1200pF (MMFD)	1.2nF		0.0000012uF / MFD	1.2pF (MMFD)	0.0012nF
0.001uF / MFD	1000pF (MMFD)	1nF		0.000001uF / MFD	1pF (MMFD)	0.001nF

Provided you have somewhere to store everything, building up a junk box is a good idea. You never know when you will need a hard-to-find component that is no longer manufactured, but then you suddenly come across one in your junk box.

Converting values

The table is also useful for converting values. When repairing electronic circuits or following circuit diagrams (also known in some countries as schematics), you may have to convert between µF, pF, and nF before ordering a capacitor.

The junk box!

Anyone who has been involved in electronics for any length of time invariably ends up with a junk box – a collection of bits and pieces that have not been thrown away in case there is something useful that can be used to repair a faulty circuit board, or act as a source of second-hand components for a construction project. You may need the above table to identify capacitors from the junk box, though you should always test these components to ensure they still work. It is unwise to keep and reuse electrolytic capacitors; they don't age well so may break down when powered up.

7400-series List

The 7400-series is an extended family of digital integrated circuit chips. The original series was designed as TTL logic chips. Over the years, a large number of sub-families have been introduced. This table is just a short list of common 7400-series devices.

Device Number	Description
7400	Quad 2-input NAND gate
7401	Quad 2-input NAND gate; open collector outputs
7402	Quad 2-input NOR gate
7403	Quad 2-input NAND gate; open collector outputs
7404	Hex inverter
7405	Hex inverter; open collector outputs
7406	Hex inverter; open collector high voltage outputs
7407	Hex buffer; open collector high voltage outputs
7408	Quad 2-input AND gate
7409	Quad 2-input AND gate; open collector outputs
7410	Triple 3-input NAND gate
7411	Triple 3-input AND gate
7412	Triple 3-input NAND gate; open collector inputs
7413	Dual 4 input NAND Schmitt triggers
7414	Hex Schmitt-trigger inverter
7415	Triple 3-input AND gate; open collector outputs
7416	Hex Inverter/Driver; open collector 15 V outputs
7417	Hex Buffer/Driver; open collector 15 V outputs
7418	Dual 4-input NAND gate with Schmitt trigger inputs
7419	Hex Schmitt trigger Inverter
7420	Dual 4-input NAND gate
7421	Dual 4-input AND gate; open collector 15 V outputs
7422	Dual 4-input NAND gate; open collector 15 V outputs
7423	Dual 4-input NOR gate with strobe
7424	Quad 2-input NAND gate with Schmitt trigger inputs
7425	Dual 4-input NOR gate with strobe
7426	Quad 2-input NAND gate; open collector 15 V outputs

Low-power CMOS logic devices offering similar functions to the 7400-series are also available via a 4000-series.

The 4000-series CMOS chips are not pin compatible with the 7400-series chips.

4000-series List

There are many ICs in the 4000-series. For simplicity, this page only covers a small selection of the most useful gates. For each IC there is a diagram showing the pin arrangement, and brief notes explain the function of the pins where necessary.

The 4000-series is good for frequencies up to 1 MHz, but above that the 74-series is better.

Characteristics of the 4000-series

- 3 V to 15 V supply.
- Very high impedance (resistance) inputs.
- All unused inputs MUST be connected to the supply.
- Outputs can sink and source only about 1 mA.
- One output can drive up to 50 inputs (Fan-out).
- Power consumption for the IC is very low, only a few μW.
- Frequency, good up to 1 MHz.

Power consumption of the 4000-series ICs is much greater at high frequencies; e.g., from μW at low frequencies to a few mW at 1 MHz.

Quad 2-input gates

- 4001 quad 2-input NOR
- 4011 quad 2-input NAND
- 4030 quad 2-input EX-OR (now obsolete)
- 4070 quad 2-input EX-OR
- 4071 quad 2-input OR
- 4077 quad 2-input EX-NOR
- 4081 quad 2-input AND
- 4093 quad 2-input NAND with Schmitt trigger inputs

The 4093 has Schmitt trigger inputs to provide good noise immunity. They are ideal for slowly changing or noisy signals. The hysteresis is about 0.5V with a 4.5V supply and almost 2V with a 9V supply.

4001 4011 4030 4070 4071 4077 4081 4093

Function	Pin	Pin	Function
input gate 1	1	14	+3 to +15V
input gate 1	2	13	input gate 4
output gate 1	3	12	input gate 4
output gate 2	4	11	output gate 4
input gate 2	5	10	output gate 3
input gate 2	6	9	input gate 3
0V	7	8	input gate 3

Triple 3-input gates

- 4023 triple 3-input NAND
- 4025 triple 3-input NOR
- 4073 triple 3-input AND
- 4075 triple 3-input OR

Notice how gate 1 is spread across the two ends of the package.

4023 4025 4073 4075

Function	Pin	Pin	Function
input gate 1	1	14	+3 to +15V
input gate 1	2	13	input gate 3
input gate 2	3	12	input gate 3
input gate 2	4	11	input gate 3
input gate 2	5	10	output gate 3
output gate 2	6	9	output gate 1
0V	7	8	input gate 1

Dual 4-input gates

- 4002 dual 4-input NOR
- 4012 dual 4-input NAND
- 4072 dual 4-input OR
- 4082 dual 4-input AND

NC = No Connection (unused pin).

4002 4012 4072 4082

Function	Pin	Pin	Function
output gate 1	1	14	+3 to +15V
input gate 1	2	13	output gate 2
input gate 1	3	12	input gate 2
input gate 1	4	11	input gate 2
input gate 1	5	10	input gate 2
NC	6	9	input gate 2
0V	7	8	NC

15 Glossary

Glossary of Electronic Terms

A

AC coupling
A circuit that passes an AC signal while blocking any DC voltage.

Active component
A component that needs to be powered in some way to make it work, such as changing the amplitude of a signal between input and output. Transistors and integrated circuits are active components.

ADC
Abbreviation for *Analog to Digital Converter.*

Amplifier
A circuit that increases the voltage, current, or power of a signal or signals.

Amplitude
The magnitude or size of a signal voltage or current.

Analogue
Information represented as a continuously varying voltage or current, as opposed to digital data that is varying between two discrete levels.

Anode
The P-type material of a diode or the positive terminal or electrode of a device.

Attenuate
Reduce the amplitude of a signal, voltage, or current. The opposite of amplification.

Average value
The average of all the instantaneous measurements in one half cycle.

B

Bandwidth
Width of the band of frequencies that a circuit can operate at or pass. The range between the upper and lower operating cut-off points.

Base
The region that lies between the emitter and collector of a bipolar junction transistor (BJT).

Bias
A DC voltage applied to a device to control its operation.

Bipolar Junction Transistor (BJT)
Three-terminal device where emitter to collector current is controlled by the base current.

Bridge rectifier
Configuration using four diodes in a circuit to provide full wave rectification. Converts an AC voltage to a DC voltage consisting only of positive half cycles.

Buffer
An amplifier used to isolate a load from a source. An electronic circuit used to isolate the input from the output.

BW
Abbreviation for *bandwidth*.

Bypass capacitor
A capacitor used to provide an AC ground at some point in a circuit. Used to short AC signals to the ground in a way that any AC noise present on a DC signal is removed, producing a much cleaner and purer DC signal.

Capacitance
The ability of a capacitor to store an electrical charge. The basic unit of capacitance is the farad.

Capacitor
An electronic component having capacitive reactance.

Cathode
The N-type material in a junction diode. The negative terminal or electrode of a device.

Center tapped transformer
A transformer with a connection at the electrical center of a winding.

Charge
Quantity of electrical energy.

Circuit
Interconnection of individual electronic components such as resistors, transistors, capacitors, inductors, and diodes through which electric current can flow.

Circuit diagram
Illustration of an electrical or electronic circuit with the components represented by their symbols.

Clamp
A diode circuit used to change the DC level of a waveform without distorting the waveform.

Collector
The semiconductor region in a bipolar junction transistor through which a flow of charge carriers leaves the base region.

Common base amplifier
A transistor circuit in which the base connection is common to both input and output.

Common collector amplifier
A transistor circuit in which the collector connection is common to both input and output.

...cont'd

Common emitter amplifier
A transistor circuit in which the emitter connection is common to both input and output.

CMOS (Complementary Metal-Oxide Semiconductor)
CMOS logic dominates the digital industry. The power requirements and component density are significantly better than with other technologies.

Comparator
An op-amp circuit that compares two inputs and provides a DC output indicating the polarity relationship between the inputs.

Conventional current flow
Concept of current flow from the positive terminal to the negative terminal. The flow of current produced by the movement of positive charges towards the negative terminal of a source.

Coulomb (symbol C)
Unit of electric charge.

Coupling
To electronically connect two circuits so that signal will pass from one to the other.

Current
Measured in amperes, it is the flow of electrons through a conductor, also known as electron flow.

D

DAC
Abbreviation for *Digital to Analog Converter.*

Darlington pair
Amplifier consisting of two bipolar junction transistors with their collectors connected together and the emitter of one connected to the base of the other. Circuit has an extremely high current gain and input impedance.

DC
A current that flows in only one direction. Abbreviation for *Direct Current.*

Decibel (dB)
A unit used to measure the intensity of a sound or the power level of an electrical signal. A logarithmic representation of gain or loss.

Depletion layer or region
The area surrounding a P-N junction that is depleted of carriers.

Device
An electronic component or part.

Dielectric
The insulating material between two plates where an electrostatic field exists.

Dielectric strength
The maximum voltage an insulating material can withstand without breaking down.

Digital
Term relating to devices or circuits that have outputs of only two discrete levels; e.g., on or off, high or low, 0 or 1, true or false.

Diode
A two-terminal device that conducts current in one direction only.

Direct coupling
Where the output of one amplifier stage is connected directly to the input of a second amplifier or to a load. Also known as DC coupling because DC signals are not blocked.

Doping
The process of adding impurities to intrinsic (pure) silicon or germanium to improve or modify the conductivity of the semiconductor material.

E

Electric charge
Electric energy stored on the surface of a material. Also known as a static charge.

Electric field
A field or force that exists in the space between two different potentials or voltages. Also known as an electrostatic field.

Electromotive force (emf)
The force that causes the motion of electrons due to potential difference (voltage) between two points.

Electron
Smallest sub-atomic particle of negative charge that orbits the nucleus of an atom.

Electron flow
Electrical current produced by the movement of free electrons towards a positive terminal.

Electrostatic
Same as static electric charge.

Emitter
The semiconductor region from which charge carriers are injected into the base of a bipolar junction transistor.

Equivalent resistance
Total resistance of all the individual resistances in a circuit.

...cont'd

F

Farad
The basic unit of capacitance.

Feedback
A controlled portion of the output signal of an amplifier that is fed back to the input of the same amplifier.

Field Effect Transistor (FET)
A voltage-controlled transistor in which the source to drain conduction is controlled by the gate to source voltage.

Filter
Network consisting of capacitors, resistors, and/or inductors used to pass certain frequencies and block others.

Forward bias
A P-N junction bias that allows current to flow through the junction. Forward bias decreases the resistance of the depletion layer.

Frequency response
Indication of how well a circuit responds to different frequencies applied to it.

Full wave rectifier
Rectifier that makes use of both the positive and negative half cycles of a full AC waveform.

Function generator
Signal generator that can produce sine, square, triangle, and sawtooth output waveforms.

G

Gain
Increase in voltage, current, and/or power. Gain is expressed as a ratio of amplifier output value to the corresponding amplifier input value.

Ground
A conducting path between an electrical circuit or system and the earth (or some conducting body acting in place of the earth). A ground is often used as the common wiring point or reference in a circuit.

H

Half wave rectifier
A diode rectifier that converts AC to pulsating DC by eliminating either the negative or the positive alternation of each input AC cycle.

Harmonic
A signal or wave whose frequency is a multiple of the frequency of some reference signal or wave. A frequency component of a signal that is an integral multiple of the fundamental of that signal.

Hole
A gap left in the covalent bond when a valence electron gains sufficient energy to jump to the conduction band.

I

IC
Abbreviation for *Integrated Circuit.*

IC voltage regulator
Three-terminal device used to hold the output voltage of a power supply constant over a wide range of load variations.

Impedance (Z)
Measured in ohms, it is the total opposition to the flow of current offered by a circuit. Impedance consists of the vector sum of resistance and reactance.

Internal resistance
Every source has some resistance in series with the output current. When current is drawn from the source, some power is lost due to the voltage drop across the internal resistance. Usually called output impedance or output resistance.

J

Junction
Contact or connection between two or more wires or cables. The area where the P-type material and N-type material meet in a semiconductor.

Junction diode
A semiconductor diode in which the rectifying characteristics occur at a junction between the N-type and P-type semiconductor materials.

K

Kilo
Metric prefix for 1000.

Knee voltage
The voltage at which a curve joins two relatively straight portions of a characteristic curve. For a P-N junction diode, the point in the forward operating region of the characteristic curve where conduction starts to increase rapidly. For a Zener diode, the term is often used in reference to the Zener voltage rating.

L

L-C tank circuit
A circuit consisting of inductance and capacitance, capable of storing electricity over a band of frequencies continuously distributed about a single frequency at which the circuit is said to be resonant or tuned.

...cont'd

Light-Emitting Diode (LED)
A semiconductor diode that converts electric energy into electromagnetic radiation (as photons) at a visible and near infrared frequencies when its P-N junction is forward biased.

Linear
Relationship between input and output in which the output varies in direct proportion to the input.

Linear scale
A scale in which the divisions are uniformly spaced.

Load
A component or piece of equipment connected to a source that, drawing current from that source, is said to *load* the source.

Load current
Current drawn from a source by a load.

Load impedance
The vector sum of reactance and resistance in a load.

Loading effect
A large load impedance will draw a small load current and so loading of the source is small and called a light load. A small load impedance will draw a large load current from the source and is called a heavy load.

Load resistance
Resistance of a load.

M

Majority carriers
The conduction band electrons in an N-type material and the valence band holes in a P-type material. Produced by pentavalent impurities in N-type material and trivalent impurities in P-type material.

Metal Oxide Semiconductor FET (MOSFET)
A field effect transistor in which the insulating layer between the gate electrode and the channel is a metal oxide layer.

Minority carriers
The conduction band holes in N-type material and valence band electrons in P-type material. Most minority carriers are produced by temperature rather than by doping with impurities.

Monostable
A circuit that has one stable state. When actioned, the circuit will return to the stable state after a predetermined fixed amount of time.

MOSFET
Abbreviation for Metal Oxide Semiconductor Field Effect Transistor, also known as an Insulated Gate Field Effect Transistor.

Multivibrator
An electronic circuit used to implement a variety of simple two-state systems, which may be astable, bistable, or monostable; e.g., timers, flip-flops, and relaxation oscillators.

Mutual conductance
Ratio of a change in output current to the change in input voltage that caused it.

N

N-type semiconductor
A semiconductor compound formed by doping an intrinsic semiconductor with a pentavalent element. An N-type material contains an excess of conduction band electrons.

Negative
Terminal that has an excess of electrons.

Negative charge
A charge that has more electrons than protons.

Negative temperature coefficient
Term used to describe a component whose capacitance or resistance decreases when the temperature increases.

Node
Junction or branch point in a circuit.

Noise
The unwanted electromagnetic radiation within an electrical system.

Non-inverting input
The terminal on an operational amplifier that is identified by a plus sign.

Non-linear scale
A scale in which the divisions are not equally spaced; also called logarithmic.

NPN transistor
A bipolar junction transistor in which a P-type base element is sandwiched between an N-type emitter and an N-type collector.

O

One Shot – monostable circuit that produces just one single pulse when triggered.

...cont'd

P

P-type semiconductor
A semiconductor compound formed by doping an intrinsic semiconductor with a pentavalent element. A P-type material contains an excess of holes.

Passive component
Component that does not amplify a signal. Resistors, capacitors, and inductors are examples.

Peak to peak
Difference between the maximum positive and maximum negative values of an AC waveform.

PNP transistor
A bipolar junction transistor with an N-type base and P-type emitter and P-type collector.

Positive feedback
A feedback signal that is in phase with an amplifier input signal. Positive feedback is necessary for oscillation to occur.

Potential difference
Voltage difference between two points that will cause current to flow in a closed circuit.

Potentiometer
A variable resistor with three terminals. The mechanical turning of a shaft can be used to produce variable resistance and potential. A volume control is an example of a potentiometer.

Q

Quiescent point (Q point)
A point on the DC load line of a given amplifier that represents the quiescent (no signal) value of output voltage and current for the circuit.

R

Rectification
The process of converting AC to DC.

Rectifier
Diode circuit that converts Alternating Current into (pulsating) Direct Current.

Regulated power supply
Power supply that maintains a constant output voltage under changing load conditions.

Regulator
Device or circuit that maintains a desired output under changing conditions.

Resistance (symbol R)
Opposition to current flow, dissipation of energy in the form of heat, measured in ohms (Ω).

Resistor
Component made of material that opposes flow of current and therefore has some value of resistance.

Reverse bias
Bias on a P-N junction that allows only leakage current (minority carriers) to flow. Positive polarity on the N-type material and negative polarity to the P-type material.

Reverse breakdown voltage
Amount of reverse bias that will cause a P-N junction to break down and conduct in the reverse direction.

S

Schematic diagram
Illustration of an electrical or electronic circuit with the components represented by their symbols.

SDRAM
Synchronous dynamic random-access memory, the most popular form of digital memory today. It differs from previous-generation DRAM in that all signal timing is relative to one clock.

Semiconductor
An element that is neither a good insulator nor a good conductor, but lies somewhere between the two. Characterized by a valence shell containing four electrons, the most frequently used semiconductors in electronics are germanium and silicon.

Series circuit
Circuit in which the components are connected end to end so that current has only one path to follow through the circuit.

Signal to noise ratio (SNR)
Ratio of the magnitude of the signal to the magnitude of noise, usually expressed in decibels.

Silicon (Si)
Non-metallic element (atomic number 14) used in pure form as a semiconductor.

Solid state
Refers to circuits where signals pass through solid semiconductor material such as transistors and diodes, as opposed to valves or vacuum tubes where signals pass through a vacuum.

Spectrum analyzer
Instrument used to display the frequency domain of a waveform, plotting amplitude against frequency.

...cont'd

T

Transistor
Derived from **trans**fer res**istor**. Semiconductor device that can be used as an amplifier or an electronic switch.

U

USB (Universal Serial Bus)
An interface for connecting peripherals, including test instruments, to computers.

V

Varactor diode
P-N junction diode with a high junction capacitance when reverse biased. Most often used as a voltage-controlled capacitor. The varactor is also called varicap, tuning diode, and epicap.

Variable capacitor
Capacitor whose capacitance can be changed by varying the effective area of the plates or the distance between the plates.

Variable resistor
Resistor whose resistance can be changed by turning a shaft or moving a wiper.

Volt
Unit of potential difference or electromotive force. One volt is the potential difference needed to produce one ampere of current through a resistance of one ohm.

Voltage (V)
Term used to designate electrical pressure or force that causes current to flow.

Voltage divider
Fixed or variable series resistor network connected across a voltage to obtain a desired fraction of that voltage.

Voltage drop
Voltage or difference in potential developed across a component due to current flow.

Voltage regulator
Device or circuit that maintains constant output voltage within certain limits despite changes of line voltage and/or load current.

W

Watt
A unit of power.

Z

Zener diode
Semiconductor diode in which reverse breakdown voltage current causes the diode to develop a constant voltage. Used as a clamp for voltage regulation.

Index

Symbols

A

B

C

D

E

F

G

H

I

J

K

L

M

N

O

P

Q

R

S

T

U

V

W

Z